PERGAMON INTERNATIONAL LIBRARY
of Science, Technology, Engineering and Social Studies
The 1000-volume original paperback library in aid of education, industrial training and the enjoyment of leisure
Publisher: Robert Maxwell, M.C.

FAST BREEDER REACTORS

An Engineering Introduction

PERGAMON BOOKS OF RELATED INTEREST

CARTER et al.
Management of Low-Level Radioactive Waste (3 volumes)

CEC
Fusion Technology 1980

CHICKEN
Nuclear Power Hazard Control Policy

EL-HINNAWI
Nuclear Energy and the Environment

GRENON
The Nuclear Apple and the Solar Orange

HUNT
Fission, Fusion and the Energy Crisis, 2nd Edition

KEMENY
Report of the President's Commission on the Accident at Three Mile Island

LEWINS
Nuclear Reactor Kinetics and Control

MURRAY
Nuclear Energy, 2nd Edition

PETROSY'ANTS
Problems of Nuclear Science and Technology

SILVENNOINEN
Reactor Core Fuel Management

SILVERMAN
Energy Storage

SHABALIN
Fast Pulsed and Burst Reactors

PERGAMON RELATED JOURNALS

(Free Specimen Copies Gladly Sent on Request)

ANNALS OF NUCLEAR ENERGY
ENERGY
ENERGY CONVERSION AND MANAGEMENT
INTERNATIONAL JOURNAL OF HYDROGEN ENERGY
PLASMA PHYSICS
POWER STATION PRACTICE (USSR)
PROGRESS IN NUCLEAR ENERGY
SOLAR ENERGY

FAST BREEDER REACTORS

An Engineering Introduction

by

A. M. JUDD

M.A., M.Inst. P.

Dounreay Nuclear Power Development Establishment
United Kingdom Atomic Energy Authority
Caithness, Scotland

PERGAMON PRESS

OXFORD · NEW YORK · TORONTO · SYDNEY · PARIS · FRANKFURT

U.K.	Pergamon Press Ltd., Headington Hill Hall, Oxford OX3 0BW, England
U.S.A.	Pergamon Press Inc., Maxwell House, Fairview Park, Elmsford, New York 10523, U.S.A.
CANADA	Pergamon Press Canada Ltd., Suite 104, 150 Consumers Rd., Willowdale, Ontario M2J 1P9, Canada
AUSTRALIA	Pergamon Press (Aust.) Pty. Ltd., P.O. Box 544, Potts Point, N.S.W. 2011, Australia
FRANCE	Pergamon Press SARL, 24 rue des Ecoles, 75240 Paris, Cedex 05, France
FEDERAL REPUBLIC OF GERMANY	Pergamon Press GmbH, 6242 Kronberg-Taunus, Hammerweg 6, Federal Republic of Germany

First edition 1981

British Library Cataloguing in Publication Data

Judd, A. M.
Fast breeder reactors. - (Pergamon international library).
1. Breeder reactors
I. Title
621.48'34 TK9203.B7 80-41929

ISBN 0-08-023220-5 (Hardcover)
ISBN 0-08-023221-3 (Flexicover)

In order to make this volume available as economically and as rapidly as possible the author's typescript has been reproduced in its original form. This method has its typographical limitations but it is hoped that they in no way distract the reader.

Preface

This book is intended for the newcomer to the study of fast breeder reactors, either as a student or at a later stage of his or her career. It will be most useful to someone who already has some knowledge of nuclear reactors. There are many excellent introductory texts for the beginner, but they all concentrate on thermal reactors. The purpose of this book is to provide an up-to-date account of fast reactors for those who want to take the next step.

It is not comprehensive. Among the references are books and review papers treating subjects such as fast reactor physics, reactor fuel, fast reactor safety, etc. in depth. I hope that they will enable the reader to find out more about the aspects of fast reactors he or she is most interested in. I have made the book brief partly to make the writing of it manageable, but also in the hope that it will be readable and coherent. A complete account of fast reactors is impossible for any single author unless he is extraordinarily talented or has a lot of spare time. The alternatives are either a book written by several authors, with all the problems of different standards and incompatible approaches, or a short book which selects what is most important and leaves out the rest. The latter is what I have attempted.

I have divided the subject matter up in Chapters according to discipline, and Chapter 1 about the physics of fast reactors is the most detailed and mathematical. This is to give those who have to use the numbers produced by the complex computer codes which predict core performance some idea of where they come from. Chapters 3 and 4 are about the application of mainly conventional engineering disciplines and so it is not necessary to explain the theory in detail. Chapter 2 is mainly about the chemistry of fast reactor fuel. It is an area where knowledge is still expanding quickly. The main points are clear, but a lot of the details remain to be filled in.

The introduction includes a brief summary of the history of fast reactors. So far it has been a story of steady development, marked by an unusual degree of international cooperation and propelled by the prospect of making accessible enormous reserves of energy. It is inevitable that in the future the cooperation will be modified by commercial rivalry as the benefits become more immediately profitable. The greatest danger, however, is that the benefits might be lost because of unfounded fears about safety. In Chapter 5 I have tried to show how safety can be attained by careful attention to detail in design. The same care is needed in all parts of the fast reactor industry - in construction, operation, maintenance and inspection -

and requires people who are professionally highly competent. I hope this book will help them in their work and so assist the development of fast reactors to help fulfil the world's need for energy.

Many of my colleagues in the atomic energy industry have been very generous in helping me to write this book, and they are for too numerous to mention them all by name. There are some, however, whose contribution has been so great that I must acknowledge them. These are Mr. Ken Brindley of the Nuclear Power Company, and Mr. Colin Gregory, Dr. Ken Swanson and Dr. Derek Smith of the Atomic Energy Authority. I am also very grateful to Mrs. Ruth James for many hours of invaluable editorial help, and to my wife who did the typing.

Thurso, September 1980.

Contents

Introduction

Chain Reactions

Early in 1939 Meitner and Frisch suggested that the correct interpretation of the results observed when uranium is bombarded with neutrons is that the uranium nuclei undergo fission. This started a rush of research activity and within a few months two very important things became clear: that fission releases a large amount of energy; and that fission of a nucleus by one neutron liberates usually two or three new neutrons. It was soon realised that these discoveries disclosed the possibility of a chain reaction which would produce power.

There is a serious difficulty, however, in making a chain reaction work. Natural uranium consists of two isotopes: ^{235}U (with an abundance of 0.7%) and ^{238}U (99.3%). Of the two ^{235}U is "fissile", which means that fission can be induced by bombarding it with neutrons of any energy. On the other hand ^{238}U undergoes fission only if the bombarding neutrons have an energy greater than about 1.5 MeV, and even then they are more likely to be captured or scattered inelastically.

Figure 1 shows the fission cross-sections of ^{235}U and ^{238}U as functions of neutron energy. In natural uranium, where the high fission cross-section of ^{235}U compensates for its low abundance, the probability of fission outweighs that of capture for neutrons with energy less than about 0.1 eV.

Neutrons generated in fission have average energies of about 2 MeV, and cannot sustain a chain reaction in natural uranium. If a neutron survives many scattering interactions, however, its kinetic energy decreases until it is in thermal equilibrium with the atoms by which it is being scattered. It is then known as a "thermal" neutron, and at room temperature its most probable energy is about 0.025 eV.

If a chain reaction is to take place, therefore, either the fission neutrons have to be reduced in energy to round about the thermal level, in which case natural uranium can be used, or the proportion of ^{235}U has to be increased substantially. Both of these routes were followed in the early work on nuclear reactors. The first led to the development of "thermal" reactors and the second to "fast" reactors, so called because the neutrons causing fission are "fast" as opposed to "thermal".

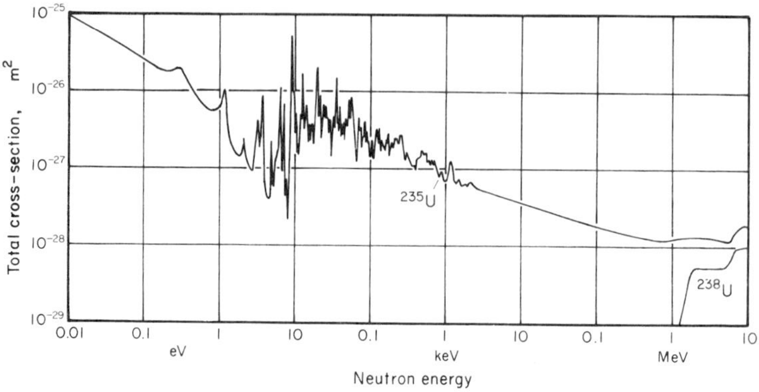

Fig. 1. Fission cross-section of ^{235}U and ^{238}U.

Breeding

A neutron captured in ^{238}U is not entirely wasted. The ^{239}U nucleus formed decays in the following way:

$$^{239}U \xrightarrow[23.5\text{ min}]{\beta^-} {}^{239}Np \xrightarrow[2.35\text{ d}]{\beta^-} {}^{239}Pu \xrightarrow[24360\text{ y}]{\alpha} {}^{235}U \text{ etc.}$$

The times shown are the half-lives for the decay processes. As far as reactor operation is concerned the long-lived plutonium isotope ^{239}Pu is the end-product of the chain.

^{239}Pu has nuclear properties quite similar to those of ^{235}U, and it can be fissioned by neutrons of all energies. In this way neutron capture provides a route for converting ^{238}U into fissile material, and so ^{238}U is known as a "fertile" isotope. ^{232}Th, which is the only naturally-occurring isotope of thorium, is also fertile. It behaves very similarly to ^{238}U: the ^{233}Th formed on capture of a neutron decays by the chain

$$^{233}Th \xrightarrow[22.1\text{ min}]{\beta^-} {}^{233}Pa \xrightarrow[27.4\text{ d}]{\beta^-} {}^{233}U \xrightarrow[162000\text{ y}]{\alpha} {}^{229}Th \text{ etc.}$$

This means that we have two naturally-occurring fertile isotopes, ^{232}Th and ^{238}U, and three fissile isotopes, ^{233}U, ^{235}U and ^{239}Pu. There are in fact other fissile and fertile isotopes as we shall see, but these five are the most important.

This ability to convert fertile isotopes to fissile raises the possibility of "breeding" new fissile material but this can be done only if enough neutrons are available. The average number of neutrons liberated in a fission is denoted by $\bar{\nu}$. Its value depends on which isotope is being fissioned and on the energy of the neutron causing the fission, but in most cases it is about 2.5. We have seen that the fact that $\bar{\nu}$ is greater than one makes a chain reaction possible: the fact that it is greater than two is almost equally important. If we have a reactor in which on average one neutron from each fission causes another fission to maintain the chain reaction, and in addition more than one of the other neutrons from the first fission is captured in fertile material, then the total number of fissile nuclei will increase as the reactor operates. This is called a "breeder" reactor.

It is sometimes said that a breeder reactor generates more fuel than it consumes. This is a rather misleading claim. It produces more fissile material than it consumes, but to do this it depends on a supply of fuel in the form of fertile material.

Although $\bar{\nu} > 2$ suggests the possibility of a breeder reactor the requirement for breeding to take place is more complicated. When a neutron interacts with a fissile nucleus it does not necessarily cause fission. It may be captured, and if it is, it is effectively lost. The important quantity in determining whether breeding is possible is the average number of neutrons generated per neutron absorbed. This is denoted by η, and it can be seen that

$$\eta = \bar{\nu}\ \sigma_f/(\sigma_f + \sigma_c).$$

η is a function of the neutron energy E, and its variation with E for the three fissile isotopes is shown in Fig. 2.

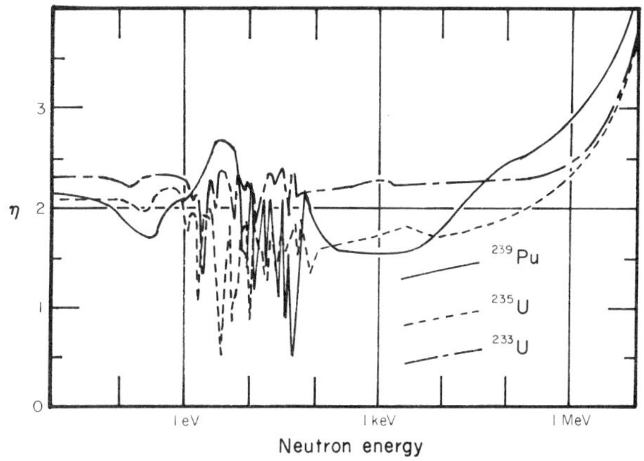

Fig. 2. η for ^{233}U, ^{235}U and ^{239}Pu.

Of these η neutrons one is needed to maintain the chain reaction. Some of the remainder are lost either because they diffuse out of the reactor or because they are captured by some of the other materials, such as the coolant or the reactor structure, which are inevitably present. The others can be captured by fertile nuclei to create fissile nuclei. If we denote the number of neutrons lost per neutron absorbed in fissile material by L and the number captured in fertile material by C, then C is the number of fissile nuclei produced per fissile nucleus destroyed and is given by

$$C \simeq \eta - 1 - L.$$

(This is only a rough value because there are other things which may happen to neutrons, but it is adequate for now. An accurate treatment has to include the effect of fast fission in the fertile nuclei, and is given in Chapter 1.)

If C is greater than one, as it must be if the reactor is to breed, it is known as the "breeding ratio". If it is less than one it is called the "conversion ratio". There is no logical reason behind the two names for the same quantity. The usage grew up as different words were used in the contexts of different reactor systems.

In practice L cannot be reduced below about 0.2, so that breeding is possible only if η is greater than about 2.2. Figure 2 shows how this can be brought about. A fast reactor using any of the three fissile materials can be made to breed although ^{239}Pu gives the widest margin and ^{235}U will allow breeding only if the energy of the neutrons causing fission is not allowed to fall much below 1 MeV. In all cases the higher the neutron energy the better the breeding ratio. A thermal reactor using ^{233}U may just be able to breed but the margin is very slender.

Other thermal reactors cannot breed but they can still convert a significant amount of fertile material into fissile. Conversion ratios for ^{235}U-fuelled thermal reactors are in the range 0.6 (for light water reactors) to 0.8 (for heavy water and gas-cooled reactors). L is particularly large in light water reactors because neutrons are readily absorbed by hydrogen.

Currently the most widely favoured breeder system is based on the use of ^{238}U and ^{239}Pu in fast reactors. Several experimental and prototype reactors have been built and commercial plants are under construction. We shall concentrate almost entirely on this system. As an alternative, thermal breeder reactors using ^{232}Th and ^{233}U have been considered, but the difficulties are very great and at present there are no plans for exploitation.

Energy Resources

So far we have been considering how it comes about that fast breeder reactors are possible. It is equally important to understand why they are considered to be necessary.

The most important reason for building fast breeder reactors is that they can make much better use of available energy reserves than other systems. Consider a uranium-fuelled reactor in which N atoms of ^{235}U are fissioned. While this is happening CN new fissile atoms of ^{239}Pu can be produced. If these in turn are fissioned in the same reactor and the conversion or breeding ratio C is unchanged (this is unlikely to be quite true as the fissile material has been changed but the effect on the argument is not important), a further C^2N fissile atoms are produced. If these are fissioned, C^3N are produced, and so on indefinitely. The total number of atoms fissioned is therefore $N(1 + C + C^2 + ----)$. If $C < 1$, the series converges and its sum is $N/(1 - C)$.

If the fuel is natural uranium N cannot exceed 0.7% of the total number of uranium atoms supplied. If the reactor is a thermal reactor with a conversion ratio of 0.7 and the plutonium bred is recycled indefinitely the total number of atoms fissioned cannot exceed $0.7/(1 - 0.7) \simeq 2.4\%$ of the number of uranium atoms supplied.

In a real system not even this number can be fissioned. When the fuel is processed to remove the fission products and the excess ^{238}U some ^{235}U is inevitably lost. In addition some ^{239}Pu is lost by conversion to higher isotopes of plutonium. As a result thermal convertor reactors can make use of at most about 2% of natural uranium.

For a breeder reactor, however, with $C > 1$, the series diverges and in principle all the fertile atoms supplied can be fissioned. In practice some are lost for the reasons mentioned above and the limit is about 50% of the fertile feed. Thus from a given quantity of natural uranium fast breeder reactors can fission about 25 times as many atoms as thermal convertors and as a result can extract about 25 times as much energy.

In order to determine the importance of this difference we have to have some idea of how much uranium and thorium are available. The estimation of mineral resources is a very complex matter, which we do not propose to go into here. The figures given below as illustrations are approximately correct but they should not be relied upon.

The amount of uranium available depends on the price we are prepared to pay to extract it. For $70 per kg some 1.5×10^6 tonnes of "yellowcake" (which is a concentrate containing 70 - 90% U_3O_8) are probably available. If 2% of these uranium atoms were fissioned about 1.5×10^{21} J of heat would be available. (10^{21}J is approximately 10^{18} B.Th.U, and for convenience is sometimes called 1 Q.) In comparison, it is estimated that the world's total oil reserves would give about 12 Q, and the total worldwide consumption of energy is about 0.3 Q per year at present.

Breeder reactors, however, could extract about 40 Q from this 1.5×10^6 tonnes of yellowcake, considerably more than all the world's oil could give. But the higher utilisation has another important effect. If more energy can be extracted from the fuel more can be paid for the ore. If the price were to rise to $500 per kg between 10 and 100 times as much uranium might be available from which 500 Q or more could be extracted. This can be compared with the total estimated reserves of coal which would yield about 300 Q. Even this is not the limit for at still higher prices it might be worthwhile to extract uranium from seawater, in which a total equivalent to about 10^5 Q is present.

Without breeding nuclear reactors can make an important contribution to the world's energy requirements but they cannot alleviate the long-term shortage and most of the energy reserve in ^{238}U will be wasted. Breeders, on the other hand, make available enough energy to change the long-term prospect radically.

This is not to say that we are certain to have to rely on breeder reactors for the bulk of our future energy requirements. There are possible dangers in doing so, as we shall discuss in later chapters, and there may be alternative sources of abundant energy. For example nuclear fusion of all the deuterium in the sea using the D-D reaction would yield some 10^{10} Q, but the technology for doing this is not yet proved and the schemes under consideration for utilisation of fusion are severely limited by the lack of availability of other materials. There are also renewable sources of energy depending on the flow of radiative energy from the sun to the earth which totals about 5000 Q per year. Until these sources are available for the large-scale generation of electricity, however, breeder reactors are likely to be an important way of meeting our future requirements.

Early Developments

The history of fast breeder reactors is quite dissimilar from that of thermal reactors. From the earliest days after the second world war different countries have pursued the development of different types of thermal reactor. Light water reactors have been predominant in the U.S.A., heavy water reactors in Canada, and gas-cooled reactors in the U.K. Only now in the late 1970s are there prospects of the various nationally-based lines of development converging.

In contrast virtually the same path has been followed in all the countries where work on fast reactors has been done. The reason for this seems to have been that until the 1960s at least fast reactors were seen to be commercially valuable only well in the future, so that the advantages of cooperation appeared to outweigh the disadvantages of aiding possible competitors. Thermal reactors on the other hand were commercially important from the start and were developed in competition which restricted the exchange of ideas and allowed different concepts to flourish.

International cooperation in fast reactor development was important for two main reasons. Firstly the nuclear data on which designs had to be based were inadequate until the 1960s. There was a lot to be gained from worldwide cooperation in measuring neutron cross-sections to the required accuracy and exchanging and comparing the results. Secondly cooperation to ensure the safety of fast reactors has been possible even when competition is necessary in other areas.

This need to exchange information resulted in, among other things, a series of international conferences on fast reactors which were addressed mainly to the problems of reactor physics and safety and were held in the U.S.A. and various European countries throughout the 1960s and 1970s. These, together with the continual publication of information in the scientific press, kept the thinking in different countries from diverging and encouraged parallel development.

In one respect it is not altogether certain that this has been an advantage. The use of liquid metals as coolants seems to have acquired a great deal of momentum, mainly because "everyone does it", and the search for alternatives has been discouraged. It is apparent that helium has certain advantages as a coolant, but no group has been able to develop a helium-cooled fast reactor to the point where it can be assessed fairly in comparison with a liquid-metal-cooled reactor.

The Era of Metal Fuel

Before about 1960 it was thought that a high breeding ratio was the most important quality of a fast reactor. To achieve this the mean energy of the neutrons has to be kept high, and this requires that extraneous materials, especially moderators, should be excluded where possible from the reactor core. As a result the early reactors had metal fuel, the metal being either enriched uranium or plutonium, alloyed in some cases with molybdenum to stabilise it to allow operation at higher temperature.

The critical masses of these reactors were small and the cores were correspondingly small so that for high-power operation they had to be cooled by a high-density coolant (to avoid impossibly high coolant velocities). Hydrogenous substances were precluded because hydrogen is a moderator, and so liquid metals were used. In most cases the coolant was sodium or sodium-potassium alloy. Some early experimental reactors were cooled with mercury but it is not favoured now because of its toxicity, cost, and low boiling point.

Many neutrons leaked from the small cores of these reactors and were absorbed in surrounding regions of natural or depleted uranium where the majority of the breeding took place. These were known as breeders, or blankets.

The first generation of low power experimental fast reactors were built in the late 1940s and early 1950s to demonstrate the principle of breeding and to obtain nuclear data. They included CLEMENTINE and EBR-I in the U.S.A., BR-1 and BR-2 in the U.S.S.R., and ZEPHYR and ZEUS in the U.K. CLEMENTINE, ZEPHYR, and BR-1 and 2 used plutonium fuel which in the early years was more readily available than highly-enriched uranium. Apart from ZEUS, which was a zero-power mock-up of the later D.F.R., they all had very small cores, the largest being EBR-I (6 litres), which was a small power reactor with an output of 1.2 MW.

When it came to higher powers, however, the volume of the core had to be increased to keep the heat fluxes down to reasonable levels and to allow for the extra coolant flow. The result was EBR-II and EFFBR (the Enrico Fermi Fast Breeder Reactor) in the U.S.A. and DFR (the Dounreay Fast Reactor) in the U.K. When they were designed these three were seen as prototypes of the reactors to be used in power stations, but as they were built it began to be recognised that they would be the end-point of the development of metal-fuelled fast reactors, and the principal use to which EBR-II and DFR were put was to test oxide fuel for the next generation.

The Importance of Burnup

About 1960 it became clear that there is more to a profitable fast reactor than a high breeding ratio. The fuel itself is expensive because of the original cost of the fissile material and the cost of fabricating it into fuel elements and reprocessing it after it has been irradiated in the reactor.

Fuel cannot remain in the reactor core indefinitely for a number of reasons. As irradiation proceeds the fissile material is used up. This is offset to some extent by the breeding of new fissile material but as most of this takes place in the breeder rather than the core there is a nett loss of fissile material from the core and the reactivity declines. At the same time fission products are formed. These have a small effect on reactivity in a fast reactor (unlike a thermal reactor in which they absorb the low-energy neutrons), but they disrupt the crystal structure of the fuel material and cause it to swell, and they may corrode the cladding material. In addition, the cladding is weakened by the fast neutrons so that in the end it loses its integrity and allows radioactive material to escape into the coolant.

When the fuel has been irradiated in the reactor for as long as it can stand it has to be removed and stored for a period to allow the most intense fission-product radioactivity to decay. It is then reprocessed chemically to remove the fission products and replace the fissile material, re-fashioned into fresh fuel elements, and returned to the reactor to be used again. The more frequently this has to be done the greater the cost, both because more reprocessing has to be paid for and also because more of the expensive fuel is lying unused waiting for reprocessing or being reprocessed.

The amount of irradiation the fuel can stand before it has to be removed and reprocessed is known as the "burnup". It can be measured by determining either the fraction of the total number of atoms of uranium and plutonium (or thorium and uranium if that cycle is being used) which are fissioned, or by determining the total amount of heat transferred from the fuel. The two are equivalent because one fission releases approximately the same amount of energy whatever isotope is fissioned, and because whatever its chemical form the mass of the fuel material is almost entirely made up by the mass of the uranium, plutonium or thorium atoms. It so happens that fission of all the uranium and plutonium (i.e. all the "heavy atoms") would, if it were possible, release about 10^6 MW days per tonne of fuel.

Experience with the early reactors showed that metal fuel can stand no more than about 1% burnup, or 10,000 MWd/tonne, whereas fuel in the form of dioxides (either UO_2, PuO_2, or a mixture of the two) can stand much higher burnup, up to 10% or more. It became obvious that the reprocessing and inventory costs of fuel which could stand only 1% burnup would be prohibitively high.

Metal fuel also has the disadvantage that it cannot tolerate operation at high temperature. Apart from phase-changes in the crystal structure of the metal itself, it is difficult to find a cladding material which is compatible with both fuel and coolant at temperatures above about 250°C. This severely limits the thermodynamic efficiency in converting heat to work and so restricts the electrical power output.

For these reasons metal fuel was eventually rejected in favour of oxide fuel.

The Present Day

Oxide, however, is far from perfect as a fuel material. The oxygen acts as a partial moderator, reducing the mean neutron energy and therefore decreasing the breeding ratio. The low thermal conductivity of the oxide is another disadvantage because it means that the fuel elements have to be very slender and the cost of manufacturing them is high.

There are on the other hand compensating advantages apart from the ability of oxide to withstand high burnup. The low mean neutron energy means that there are enough neutrons in the ^{238}U resonance region to make the Doppler effect important. This results in a prompt negative temperature coefficient of reactivity, which in some cases offsets the positive sodium temperature coefficient due to the large size of the core of a high-power reactor. These effects are described in Chapter 1.

Oxide fuel can be operated at a higher temperature than metal, and it can be clad in stainless steel which is cheaper than the refractory metals used in metal-fuelled reactors. Above all there is a lot of experience about it as it is widely used as a fuel for thermal reactors.

This last is the reason why oxide is preferred to carbide. A mixture of UC and PuC would have important advantages (higher thermal conductivity, higher density, less moderating effect), but because it is not so well understood caution at present dictates that oxide is to be preferred. It is possible that carbide fuels will be used widely at some time in the future.

Other fuels have been tried. Experiments were done on cermets, consisting of a sintered mixture of stainless steel and oxide. This had the advantage of high thermal conductivity and a porous structure which could accommodate fission products. It was rejected because absorption of neutrons in the steel was very detrimental to the breeding ratio. Molten fuel was tried in the LAMPRE experiment. This reactor had molten plutonium fuel clad in tantalum. The problems of corrosion and of the accumulation of bubbles of fission product gas in the molten fuel were so severe that the line of development was not pursued.

So mixed oxide fuel, stainless steel cladding and structure, and sodium coolant, have become accepted almost universally as the route for the development of fast breeder reactors. These materials restrict the designer's choice of variables such as the dimensions of the fuel elements and the core and as a result all current fast breeder reactors, from whatever country, show marked similarities. The British Prototype Fast Reactor (PFR), French Phenix, Russian BN-350, German SNR-300, Japanese MONJU, and the proposed Clinch River Breeder Reactor (CRBR) in the U.S.A. are, as far as the design in the reactor core is concerned, very much alike. They are all prototypes to be followed by full-scale production reactors for commercial generation of electricity such as Super-Phenix in France, the Commercial Fast Reactor (CFR) in Britain, and BN-600 in Russia.

Because of this remarkable international consensus on the design of fast breeder reactors what follows is concentrated on reactors of this type. We shall refer at times to some of the possible alternatives like carbide fuels or gas coolants, but we shall deal in the main with the oxide-fuelled, sodium-cooled reactors which, for the next decade at least, will dominate the field.

CHAPTER 1

Physics

1.1 INTRODUCTION

1.1.1 Physics and Design

Reactor physics is the study of the nuclear reactions that take place in a reactor, and it is fundamental to design in two main ways. Firstly, criticality is a question of reactor physics. A designer has to determine the size and composition needed to make the reactor critical, he has to predict the effect on reactivity of movement of the control rods and burnup of the fuel, and he has to estimate the reactivity changes which come about in the course of normal operation and under abnormal conditions. Secondly, he has to know the rate at which various nuclear reactions take place for on these depend the power generated and its distribution within the reactor, the burnup of the fuel, the breeding of new fuel, the alteration of the properties of the materials in the reactor, the build-up of radioactivity, and the need for radiation shielding.

Reactor physics is not, however, the only important influence on design. Heat transfer, structural, metallurgical, and safety considerations are also important, and the design ultimately chosen is a compromise. In reaching this compromise a designer's overriding aims are that the reactor should be as profitable as possible, provided that it is safe.

1.1.2 Typical Reactors

The physics of a fast reactor depends on the composition of the core and on its size and shape but not strongly, as explained in section 1.1.3, on the details of the structure of individual fuel elements. The composition is determined as follows. The only material in the core which is essential to the physics is the fuel (unlike the case of a thermal reactor in which the moderator is also essential) but the demands of heat transfer nevertheless require that about 50% of the core volume is occupied by coolant. Structural material takes up another 20% or so leaving about 30% for fuel.

To minimise neutron leakage from the core the best shape is a sphere, but this is very inconvenient. A right circular cylinder with a vertical axis and with the coolant flowing upwards through it is much simpler. The various demands of heat transfer and fluid mechanics, which are discussed in Chapter 3, limit the height of the core to about 1 m and the average power density to about 500 MW m^{-3}, so that the core has to be a squat cylinder, and the larger the reactor power the larger its diameter. As a result more neutrons leak out than from a spherical core so that the critical mass is increased and the breeding reduced.

1.1.3 Comparison with Thermal Reactors

Before describing the physics of fast reactors it may be worthwhile pointing out the main differences between fast and thermal reactors.

The most important is that the composition of the fuel is different. In a fast reactor the fraction of fissile material is about 20% compared with 0.7 - 3% in a thermal reactor. This and the lack of a moderator means that fast reactor cores are much smaller (with dimensions of the order of 1 m compared with 3 m for light water reactors and 10 m for graphite or heavy water reactors) and the power density is much higher.

In a fast reactor thermal neutrons are almost absent so that the materials with high thermal neutron absorption cross-sections, which are so important in thermal reactors, do not matter nearly as much in a fast reactor. Fission products such as ^{135}Xe and ^{147}Sm and impurities such as boron are relatively insignificant. There is no xenon poisoning problem for a fast reactor and the decrease of reactivity with burnup of the fuel due to the accumulation of fission products is much slower than in a thermal reactor. Because most materials have similar cross-sections for fast neutrons, nuclear considerations place much less severe limits on the choice of materials for a fast reactor core.

The mean free path of fast neutrons is longer than that of thermal neutrons, so that the core of a fast reactor is more closely coupled than that of a thermal reactor. There is no question of zonal instability, there is no significant depression of the neutron flux in the fuel elements and except for some experimental reactors there is no need for cell calculations to predict the effects of heterogeneity.

The temperature coefficients of reactivity come from entirely different sources (the Doppler effect and coolant expansion in fast reactors rather than moderator expansion and change in thermal energy), but the magnitudes are similar so the dynamics of fast and thermal reactors are very similar in normal operation. Only in very rapid transients when the prompt neutron lifetime (of the order of 10^{-7} s in a fast reactor, 10^{-3} s in a thermal reactor) is important is there any difference.

In spite of the simplicity of a fast reactor neutron flux calculations are much more complex because the simplifying assumptions valid for a thermal reactor cannot be made. In a thermal reactor most of the neutrons have energies in a narrow range and one-group or few-group calculations are useful. In a fast reactor the neutrons have a wide range of energies and multigroup calculations are essential. There is no fast reactor equivalent to the "four-factor formula".

1.2 CALCULATION METHODS

1.2.1 Transport Theory and Diffusion Theory

The neutron density n in a reactor at steady state is in general a function of position $\underline{r}$, the energy E of the neutrons, and the direction $\underline{\Omega}$ in which they are travelling. In symbols, we have $n = n(\underline{r}, E, \underline{\Omega})$. n obeys a linear version of the Boltzmann equation called the neutron transport equation. The form of this equation and its derivation are given in many standard works on neutron transport (e.g. Davison, 1957; Duderstadt, 1976).

Although the resulting equation is simple in principle it can only be solved numerically, partly because there are six independent scalar variables and partly because the nuclear cross-sections, which determine the rate of fission, scattering and absorption, vary in a very complex way with neutron energy. It can, however, be simplified, because for many applications it is not necessary to take account of the direction $\underline{\Omega}$, and this gives rise to the treatment known as "diffusion theory".

Formally the transport equation is integrated over all directions and various approximations are made to obtain the energy-dependent diffusion equation. This is usually written in terms of the neutron flux ϕ, defined by

$$\phi(\underline{r}, E) = v(E) \int n(\underline{r}, E, \underline{\Omega})\, d\Omega. \qquad 1.1$$

The neutron speed v is introduced for convenience because reaction rates are proportional to nv.

The diffusion equation is accurate if the flux is isotropic or at most linearly anisotropic. This is always the case except in a strongly absorbing medium or where the properties of the medium change substantially over distances comparable to the mean free path of the neutrons.

In a fast reactor neither of these limitations is important. For fast neutrons cross-sections are usually small and mean free paths are typically 10 cm or longer. The nuclear properties of fuel, coolant and structure are very different, but as the dimensions of individual fuel elements and structural members are usually a few millimetres, over distances comparable with the mean free path large regions of the reactor can be treated as homogeneous and diffusion theory can be used. Even in control rods the capture cross-sections for fast neutrons are low enough for diffusion theory to be a good approximation for most purposes. It gives rise to the worst inaccuracies at the boundary between core and breeder.

Exceptions for which transport theory has to be used are the older small fast reactors where the dimensions of the whole core are comparable with the mean free path. Some experimental reactors are assembled from various materials in the form of plates which are typically 5 cm square and may be arranged in arrays comparable in size with the mean free paths so that transport theory has to be used. Also a modified version of diffusion theory is necessary for neutron shielding calculations.

1.2.2 Multigroup Diffusion Theory

The diffusion equation describes the distribution of ϕ with respect to $\underline{r}$ and E which are both continuous variables, but it can be solved numerically only for a finite number of distinct positions and energies. The spatial variables are treated by conventional finite difference methods, but neutron energy is usually handled rather differently. The range of energy of interest (from about 10 MeV down to 0.1 eV or lower) is divided into a series of intervals $E_0 - E_1$, $E_1 - E_2$, etc., where E_0 may be 10 MeV, and $E_1 < E_0$, $E_2 < E_1$, etc. The neutrons with energies between E_g and E_{g-1} are known as group g neutrons. The flux ϕ_g in group g is given by

$$\phi_g(\underline{r}) = \int_{E_g}^{E_{g-1}} \phi(\underline{r}, E)\, dE, \qquad 1.2$$

and ϕ_g, rather than $\phi(E)$, are calculated.

It is possible to write a separate diffusion equation for each group of neutrons (Duderstadt, 1976). If there are G groups in all, the equations for a homogeneous medium are

$$- D_g \nabla^2 \phi_g + \Sigma_{rg}\phi_g = \sum_{g'=1}^{g-1} \Sigma_{sg'\to g}\phi_{g'} + \frac{1}{k}\chi_g \sum_{g'=1}^{G} \bar{\nu}_{g'}\Sigma_{fg'}\phi_{g'} \,. \qquad 1.3$$

(1) (2) (3) (4)

This is not the most general form of the multigroup diffusion equations but it is the form usually encountered in fast reactor calculations.

The terms in the equations can be understood best by thinking of them as a balance of neutrons. If each term is multiplied by a small volume dV equation g is then a balance between the rates at which group g neutrons appear in and disappear from dV.

(1) This is the nett rate at which neutrons diffuse out of dV. D_g is a diffusion coefficient, and is assumed to be independent of position within dV.

(2) This is the rate at which neutrons in group dV are lost from group g. Σ_{rg} is the "group removal cross-section". It includes the effects of absorption and both inelastic and elastic scattering. Scattering interactions which do not remove the neutron from the group (known as "in-group scattering") do not appear in Σ_{rg}.

(3) This is the rate at which neutrons are scattered into group g. $\Sigma_{sg'\to g}$ is the transfer cross-section from group g' to g. It includes both inelastic and elastic scattering, but unless the groups are narrow or hydrogen is present elastic scattering can take a neutron from one group to the next but no further. Thus elastic scattering appears usually in the $\Sigma_{sg-1\to g}$ cross-sections, while the $\Sigma_{sg'\to g}$ for $g'<g-1$ involve only inelastic scattering. The summation in term (3) goes only up to g-1 because it can be assumed that there is no up-scattering. This is not always true for a thermal reactor.

(4) This is the rate at which neutrons are born into group g by fission. $\Sigma_{fg'}$ is the fission cross-section and $\bar{\nu}_{g'}$ is the average number of neutrons born per fission which, as we have seen, depends on the energy of the neutron causing the fission. χ_g is the fraction of fission neutrons born into group g, normalised so that $\sum\chi_g = 1$.

k can be thought of in two ways. The straightforward interpretation is that it is the effective multiplication constant, k_{eff}, of the reactor. But if $k \neq 1$ the reactor cannot be steady, ϕ_g cannot be constant, and there should be a $\partial\phi_g/\partial t$ term in the equations, so that under this interpretation equations 1.3 have a meaning for k = 1 only. Alternatively k can be thought of as the highest eigenvalue of the set of equations. They can be solved to find k and the ϕ_g, although in this case the ϕ_g have no physical meaning unless k = 1. The composition of the reactor is altered to make k = 1 and when this is achieved the ϕ_g are the group fluxes in the critical reactor.

Before the equations can be solved of course values have to be given to the constants D_g, Σ_{rg}, $\Sigma_{sg'\to g}$, $\bar{\nu}_g$, $\Sigma_{fg'}$, and χ_g. All except χ_g depend on the microscopic cross-sections for the various materials in the reactor. These vary with neutron energy so the group constants involve average values over the energy range covered by each group. We shall consider now how these averages are chosen.

In principle it is possible to make the groups so narrow that the variation of each cross-section within each group is small, but in practice this cannot be done. Figure 1.1 shows the variation of one important cross-section with energy, and it is obvious that thousands of groups would be needed. But if the distribution of the group fluxes throughout the reactor is also to be determined the equations for each group may have to be solved for hundreds of points in space. This would mean solving some 10^6 simultaneous algebraic equations, which is impossible if the cost of doing the calculations is to be kept within reason. A method has to be found of doing accurate calculations with at most 20 or 30 neutron groups.

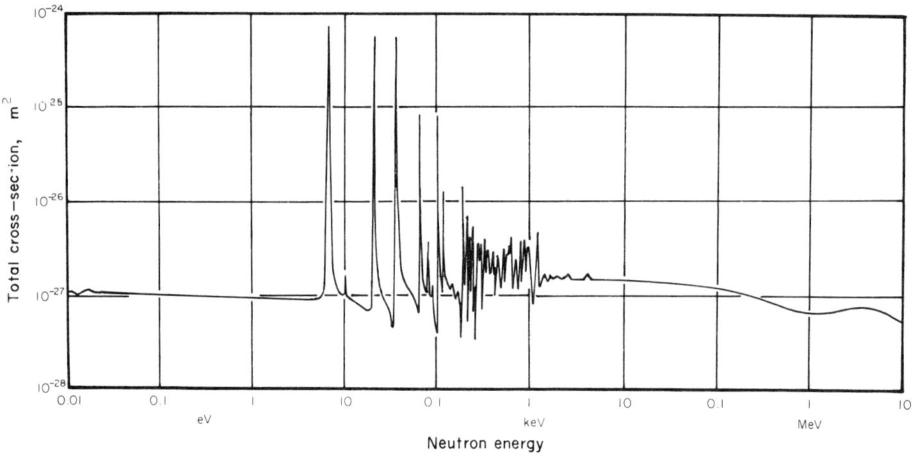

Fig. 1.1. Total cross-section of ^{238}U.

1.2.3 Fundamental Mode Calculations

The difficulty is to estimate how the flux varies within each group before the equations 1.3 are solved, and it is overcome by making a very simple assumption about the spatial variation. When this is done it is possible to estimate the details of the variation in energy (the "energy spectrum" of the flux), which can be used to calculate group constants for a proper calculation of the spatial variation. In equations 1.3 the spatial variation is contained in term (1).

The simplification is made by assuming that $\nabla^2\phi$ can be replaced by $-B^2\phi$ where B^2 is the "buckling", a constant which is the same for all energies.

The energy range is divided into a large number of fine groups so that the variation of the cross-sections within each group is small. We shall denote these fine groups by the suffix n to distinguish them from the broad groups denoted by the suffix g which are used for the spatial calculation. If B^2 is chosen suitably, corresponding to the system being critical, we can write $k = 1$. We can also normalise the fine group fluxes (as we are dealing with only one point in space) so that

$$\sum_{n=1}^{N} \nu_n \Sigma_{fn} \phi_n = 1. \qquad 1.4$$

The equation for ϕ_n then becomes

$$\phi_n = \Big(\sum_{n'=1}^{n-1} \Sigma_{sn'\to n} \phi_{n'} + \chi_n\Big)/(D_n B^2 + \Sigma_{rn}). \qquad 1.5$$

In this set of linear equations ϕ_n depends only on the $\phi_{n'}$ for $n'<n$, so they can be solved one after the other starting with ϕ_1. It is then necessary to iterate, changing the value of B^2 until equation 1.4 is satisfied.

This is the procedure of a "fundamental mode" or "normal mode" calculation. It is possible to use rather more accurate approximations to the transport equation (the "consistent P-1" or "B-1" approximations) but these are not necessary for most fast reactor applications. The important point is that this simple procedure can be carried out easily for a very large number of fine groups, typically 1000 or more, even though some iteration is involved.

If the reactor consists of several regions with different compositions a separate fundamental mode calculation has to be done for each, obtaining different neutron spectra. For breeder regions it is necessary to make $B^2 < 0$.

Once fine group spectra have been found for each region the broad group constants can be found. They are defined so that, when multiplied by the broad group flux, the resulting reaction rate is the same as would be calculated from the fine group fluxes and constants. Thus

$$\bar{\nu}_g \Sigma_{fg} = \sum_{n\epsilon g} \nu_n \Sigma_{fn} \phi_n/\phi_g, \qquad 1.6$$

where

$$\phi_g = \sum_{n\epsilon g} \phi_n. \qquad 1.7$$

The diffusion coefficient is given by

$$D_g = \sum_{n\epsilon g} D_n\phi_n/\phi_g = \sum_{n\epsilon g} (\phi_n/3\,\Sigma_{tr\,n})/\phi_g. \qquad 1.8$$

This is based on the assumption that $\nabla^2\phi$ varies with energy in the same way as ϕ. This assumption is basic to a fundamental mode calculation.

The group transfer cross-sections include contributions from both inelastic and elastic scattering. The inelastic scattering component is given by

$$\Sigma_{isg'\to g} = \sum_{n\epsilon g}\sum_{n'\epsilon g'} \Sigma_{isn'\to n}\,\phi_{n'}/\phi_{g'}, \qquad 1.9$$

but the elastic scattering has to be treated differently because if the broad groups are wide compared with $\bar{\xi}E_g$, where $\bar{\xi}$ is the mean logarithmic energy change, elastic scattering transfers neutrons from the lower end of the group only. Thus

$$\Sigma_{esg'\to g} = \sum_{n\epsilon \ell(g-1)} \Sigma_{esn}\,\phi_n/\phi_{g-1} \qquad 1.10$$

where the summation is over the fine groups roughly within $\bar{\xi}E_{g-1}$ of the group boundary.

The group removal cross-section has components due to absorption and elastic and inelastic scattering which can be treated in the same way:

$$\Sigma_{rg} = \left[\sum_{n\epsilon g}\left\{\Sigma_{an} + \sum_{n'>g}\Sigma_{isn\to n'}\right\}\phi_n + \sum_{n\epsilon \ell(g)} \Sigma_{esn}\phi_n\right]/\phi_g. \qquad 1.11$$

It is possible to make a rough estimate of the group constants without a fundamental mode calculation. In a medium with no absorption and a constant scattering cross-section $\phi(E)$ would vary as $1/E$. Though the actual spectrum departs from this over wide energy ranges over the width of a single group it is sometimes reasonably accurate to assume $\phi \propto 1/E$, and this gives for example

$$\bar{\nu}_g\,\Sigma_{fg} = \int_{E_g}^{E_{g-1}} \bar{\nu}\,\Sigma_f\, dE/EU_g \qquad 1.12$$

where

$$U_g = \ln\,(E_{g-1}/E_g). \qquad 1.13$$

Using this assumption the elastic scattering transfer cross-section is given by

$$\Sigma_{es(g-1)\to g} \simeq \bar{\xi}\,\Sigma_{es}\phi(E_{g-1})/U_{g-1}. \qquad 1.14$$

1.2.4 Resonances

The cross-sections of many nuclei exhibit sharp resonances over part of the energy range. Heavy nuclei such as uranium and plutonium have resonances in the range from about 1 eV up to 100 eV or more, as shown in Fig. 1.1 for example. In fact the apparently smoother behaviour of the cross-section above 1 keV is illusory, because there are resonances at higher energies which are so close together that it is impossible to distinguish them experimentally. Lighter elements have resonances at higher energies of 100 keV or more, although there is an isolated resonance of ^{23}Na at 3 keV.

It would be impossible to do a fundamental mode calculation for all these resonances even if they had been resolved experimentally because there would have to be far too many fine groups. Fortunately it is possible to calculate the reaction rate in a resonance directly. In outline the method is as follows.

In the vicinity of an isolated resonance the microscopic cross-section for a certain reaction, neutron capture say, is given by

$$\sigma_c = \sigma_{co}\left(\frac{E_o}{E_c}\right)^{\frac{1}{2}} \Bigg/ \left(1 + \frac{4(E_c - E_o)^2}{\Gamma^2}\right). \qquad 1.15$$

Here σ_{co} is the capture cross-section at E_o, the energy of the resonance peak, and Γ is the "width" of the resonance (the full width in energy at half the maximum cross-section). E_c is the kinetic energy of the neutron and nucleus relative to the centre of mass of the neutron-nucleus system. If the nucleus is stationary $E_c = E/(1 + 1/A)$ where A is the ratio of the mass of the nucleus to that of the neutron and E is the neutron energy.

Unless the reactor is at absolute zero temperature, however, the nucleus is unlikely to be stationary. It moves at random due to thermal agitation and this affects E_c which is increased if the nucleus happens to be moving towards the approaching neutron or decreased if it is moving away. The effect on the apparent cross-section can be calculated if the probability distribution of velocity of the nucleus is known.

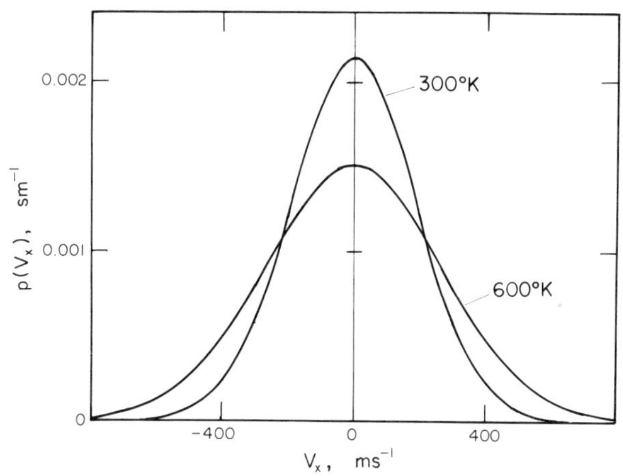

Fig. 1.2. Effect of temperature on one component of velocity for a ^{238}U nucleus.

It is usually satisfactory to assume the nucleus velocities have a Maxwell-Boltzmann distribution. The resulting distribution of the component of velocity parallel to a certain direction (which we can take as the direction of the neutron) is shown in Fig. 1.2 for atoms of ^{238}U at various temperatures.

From this the probability distribution of E_c can be deduced, and hence the mean cross-section for a neutron of energy E. This is given by

$$\bar{\sigma}_c\,(E,\,T) \simeq \sigma_{co}\left(\frac{E_o}{E}\right)^{\frac{1}{2}}\psi\,(\zeta,\,x), \tag{1.16}$$

where

$$\psi(\zeta,\,x) \equiv \frac{\zeta}{2\sqrt{\pi}}\int_{-\infty}^{\infty}\frac{\exp\;(-(x-y)^2\,\zeta^2/4)}{(1+y^2)}\,dy, \tag{1.17}$$

$$\zeta^2 \equiv \Gamma^2\;A/4E_o\;b\;T, \tag{1.18}$$

$$x \equiv 2(E - E_o)/\Gamma, \tag{1.19}$$

and b ≡ Boltzmann's constant. The approximations used in reaching equation 1.16 introduce negligible errors in most fast reactor applications.

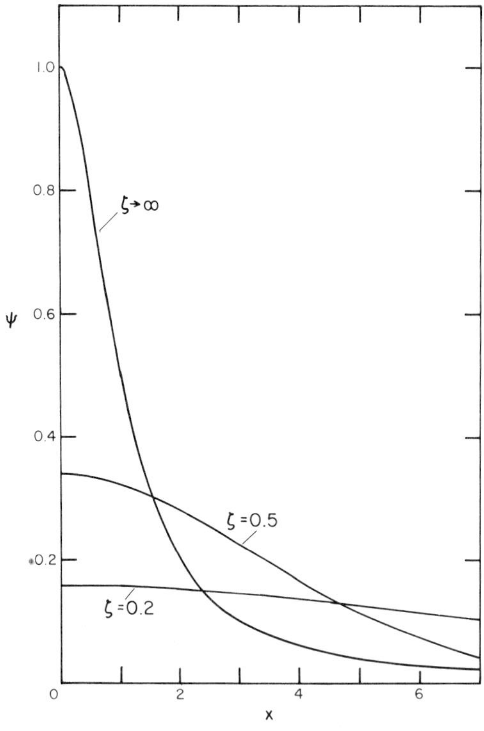

Fig. 1.3. Resonance self-shielding. The effect of temperature on effective cross-section.

Figure 1.3 shows the variation of ψ with x for various values of ζ. These curves have the shape of the effective cross-section curve in the neighbourhood of the resonance for various temperatures, higher temperature corresponding to lower ζ.

To determine the group cross-section for a neutron group containing a resonance we have to know how the flux varies in the resonance region. A simple approximation, which is fairly accurate in most cases, is the "narrow resonance approximation". Suppose we have a region made up of two materials, one of which has a constant scattering cross-section Σ_s while the other has only a single capture resonance. Suppose also that the resonance is narrow so that $\Gamma << \bar{\xi} E_o$, where $\bar{\xi}$ is the mean change in log E in neutron scattering events. It can be shown that the flux $\phi(E)$ varies as $1/E\Sigma_t$ where Σ_t is the total cross-section - i.e. $\Sigma_t = \Sigma_s + N\bar{\sigma}_c$, where N is the number of atoms of the capturing material per unit volume. The capture cross-section for a group which contains this resonance is then

$$\Sigma_{cg} = \int_{E_g}^{E_{g-1}} \frac{\bar{\sigma}_c}{\Sigma_t} \frac{dE}{E} \Bigg/ \int_{E_g}^{E_{g-1}} \frac{dE}{\Sigma_t E}, \qquad 1.20$$

and the denominator of this expression is ϕ_g, the total flux in the group. Substituting for Σ_t and making various approximations we obtain

$$\Sigma_{cg} \simeq \Gamma J/\phi_g, \qquad 1.21$$

where

$$J \equiv \frac{1}{2}\int_{-\infty}^{\infty} \frac{\psi dx}{\psi + \beta} \qquad 1.22$$

and

$$\beta \equiv \Sigma_s / N\sigma_{co}. \qquad 1.23$$

Thus the group capture cross-section and the total capture rate depend both on the temperature (via ζ) and on the scattering cross-section. The variation of J with β for various values of ζ is shown in Fig. 1.4. It can be seen that when β is large J is independent of ζ, but when β is smaller J increases as ζ decreases. This means that at "infinite dilution" when there is very little of the capturing material the capture rate is independent of temperature, but when more is present the capture rate increases as the temperature increases. This is known as the Doppler effect on the capture rate.

It happens because as the temperature increases the effective resonance becomes lower and broader. The flux at the peak of the resonance is less depressed, but this does not quite compensate for the lower cross-section so the reaction rate per unit energy ($\phi\bar{\sigma}_c$) at the resonance peak decreases. At the sides of the resonance, however, $\bar{\sigma}_c$ increases more than ϕ decreases and $\phi\bar{\sigma}_c$ increases. The increase at the sides outweighs the decrease at the peak and the total reaction rate ($\int\phi\bar{\sigma}_c\, dE$) increases.

This applies to all resonance reactions, and of particular importance in a fast reactor is the fact that both the fission rate in ^{235}U or ^{239}Pu and the absorption rate, predominantly in ^{238}U, increase with temperature. The former tends to increase reactivity and the latter to decrease it. The capture effect is greater so that the resulting temperature coefficient of reactivity, known as the Doppler coefficient, is negative, making as we shall see for stability of the reactor.

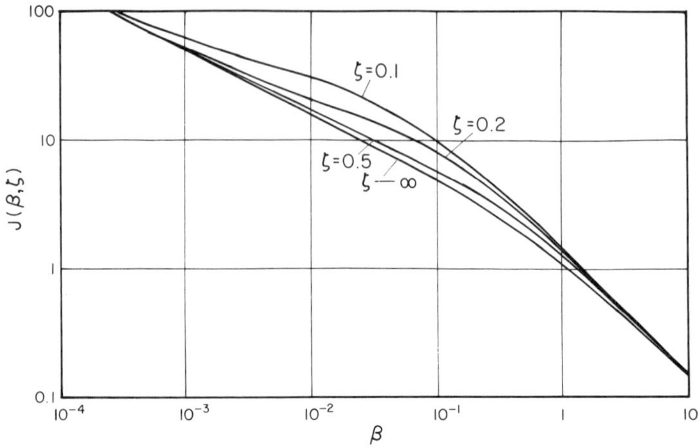

Fig. 1.4. Resonance self-shielding. The effect of dilution.

The above discussion of resonance absorption and the Doppler effect is a simple version of the whole story. Other effects which have to be taken into account are the variation of the flux integral (the denominator in equation 1.20, which is not constant), the fact that resonances in one material are not in general isolated but overlap both with each other and with those of other materials, the fact that the narrow resonance approximation is not accurate, and the existence of unresolved resonances. The way these problems can be treated is described by Hummel (1970).

1.2.5 Perturbation Theory

While it is always possible to solve equations 1.3 to find the eigenvalue k of a system it is very useful to have a means of estimating the effect of small changes. This is particularly so in the case of temperature coefficients. We have seen for example that a change in temperature makes small alterations to certain group cross-sections by means of the Doppler effect. We need a method of estimating the resultant change in k, and this is provided by Perturbation Theory.

Suppose the perturbation in which we are interested results in an increase $\delta\Sigma_{cg}$ in the capture cross-section for group g in a small region dv at a point $\underline{r}$ in the reactor. The increased rate of capture of group g neutrons will then be $\phi_g(\underline{r})\delta\Sigma_{cg}$ dv. The resulting effect on k depends on g and $\underline{r}$: neutrons in some parts of the reactor and at some energies are more important than others. Neutrons captured at the edge of the core, which were anyway quite likely to have leaked out and been lost, are less important than neutrons captured at the centre, which were more likely to cause another fission.

This distinction can be represented by assigning an importance, ϕ^*, to each neutron. ϕ^* depends on energy and position, so $\phi^* \equiv \phi^*_g(\underline{r})$. ϕ^* can be defined in a simple physical way as follows (although it should be remembered that this is not the only way of interpreting it).

If a reactor with no source of neutrons apart from fission is exactly critical and operating at a certain mean power P, and a single neutron in group g is artificially introduced into it at a point $\underline{r}$, the power will, after a time, settle down to a new mean value $P + \delta P$, where δP is a random quantity. If we repeat the experiment many times it is possible in principle to find the average change in power due to the introduction of one new neutron. This we call $\phi^*_g(\underline{r})$, the importance of the neutron in group g at $\underline{r}$.

With this definition in mind we can now see that the rate of change of power due to our increased capture $\delta\Sigma_{cg}$ at $\underline{r}$ is

$$\dot{P} = -\phi^*_g(\underline{r})\,\phi_g(\underline{r})\,\delta\Sigma_{cg}\,dv. \tag{1.24}$$

In order to determine the effect on k we have also to know what the actual power is in terms of ϕ^*. The rate at which new group g neutrons are being generated in dv is $S_g\,dv$, where

$$S_g = \sum_{g'} \chi_g\,\bar{\nu}_{g'}\,\Sigma_{fg'}\,\phi_{g'}. \tag{1.25}$$

If the mean lifetime of these neutrons is λ, the number of group g neutrons in dv is $\lambda S_g\,dv$, and their contribution to the reactor power is $\lambda\phi^*_g\,S_g\,dv$. The total reactor power is therefore

$$P = \sum_g \int_R \lambda\phi^*_g\,S_g\,dv = \int_R \sum_g \sum_{g'} \lambda\chi_g\,\bar{\nu}_{g'}\,\Sigma_{fg'}\,\phi_{g'}\,\phi^*_g\,dv, \tag{1.26}$$

and the time constant is $\tau = \dot{P}/P$.

Elementary reactor kinetics using the point kinetics model shows that τ is related to reactivity ρ by $\tau = \lambda/\rho$, where ρ is defined as $(k-1)/k$. Thus we have finally,

$$\rho = -\phi^*_g(\underline{r})\,\phi_g(\underline{r})\,\delta\Sigma_{cg}\,dv/C, \tag{1.27}$$

where

$$C \equiv \int_R \sum_g \sum_{g'} \chi_g\,\bar{\nu}_{g'}\,\Sigma_{fg'}\,\phi_{g'}\,\phi^*_g\,dv, \tag{1.28}$$

the integral being taken to run over the whole of the reactor.

This treatment glosses over a lot of difficulties, chiefly that λ is different for neutrons in different groups, but it illustrates the principle. A rigorous treatment can be found in Duderstadt (1976) for example, and shows that in the general case of perturbations to any of the group constants the change $\delta\rho$ in reactivity is given by

$$\delta\rho = \frac{1}{C}\int_R \Big\{ \sum_g \sum_{g'} \left(\delta(\chi_g\,\bar{\nu}_{g'}\,\Sigma_{fg'}) + \delta\Sigma_{sg'\to g}\right)\phi^*_g\,\phi_{g'} - \sum_g \left(\delta\Sigma_{rg}\,\phi^*_g\,\phi_g + \delta D_g \nabla\phi^*_g\cdot\nabla\phi_g\right)\Big\}\,dv \tag{1.29}$$

This complicated expression, and indeed much of multigroup diffusion theory, is expressed much more easily in matrix form. Equations 1.3 can be written

$$\underline{M}\ \underline{\phi} = \frac{1}{k}\ \underline{F}\ \underline{\phi} \qquad 1.30$$

where $\underline{\phi}$ is the column vector $\begin{bmatrix}\phi_1\\ \phi_2\\ ,\\ ,\end{bmatrix}$ and $\underline{M}$ and $\underline{F}$ are the matrices

$$\underline{M} \equiv \begin{bmatrix} -D_1\nabla^2 + \Sigma_{r1} & 0 & 0 & - \\ -\Sigma_{s1\to2} & -D_2\nabla^2 + \Sigma_{r2} & 0 & - \\ -\Sigma_{s1\to3} & -\Sigma_{s2\to3} & -D_3\nabla^2 + \Sigma_{r3} & - \\ , & , & , & \end{bmatrix}, \qquad 1.31$$

and

$$\underline{F} \equiv \begin{bmatrix} \chi_1\bar{\nu}_1\Sigma_{f1} & \chi_1\bar{\nu}_2\Sigma_{f2} & - & - \\ \chi_2\bar{\nu}_1\Sigma_{f1} & \chi_2\bar{\nu}_2\Sigma_{f2} & - & - \\ , & , & , & , \end{bmatrix}. \qquad 1.32$$

If $\underline{\phi}^* \equiv \begin{bmatrix}\phi_1^*\\ \phi_2^*\\ ,\\ ,\end{bmatrix}$ it can be shown that $\underline{\phi}^*$ is the solution of

$$\underline{M}^T\ \underline{\phi}^* = \frac{1}{k}\ \underline{F}^T\ \underline{\phi}^*, \qquad 1.33$$

where $\underline{M}^T$ and $\underline{F}^T$ are the transposes of $\underline{M}$ and $\underline{F}$ (i.e. the i, j element of $\underline{M}^T$ is the j, i element of $\underline{M}$ etc.). $\underline{M}^T$ and $\underline{F}^T$ are actually the "adjoints" of $\underline{M}$ and $\underline{F}$, but because $\underline{M}$ and $\underline{F}$ are real the adjoint (which is the transpose of the complex conjugate) is the same at the transpose. ϕ^* is often called the "adjoint flux", but this is misleading because it does not have the properties of a flux. In particular if two adjacent groups g and g+1 are combined to form a new group g',
$\phi_{g'} = \phi_g + \phi_{g+1}$, but ϕ^* is not additive and $\phi^*_{g'}$ is an average between ϕ^*_g and ϕ^*_{g+1}.

Equation 1.29, which is the basic result of first-order perturbation theory, in matrix notation is

$$\delta\rho = \int_R \underline{\phi}^*\ (\delta\underline{F} - \delta\underline{M})\ \underline{\phi}\ dv \Big/ \int_R \underline{\phi}^*\underline{F}\ \underline{\phi}\ dv. \qquad 1.34$$

This is called "first-order" for two reasons. The first is that second-order small quantities have been neglected in the usual way. The second is that $\underline{\phi}$ and $\underline{\phi}^*$ both refer to the unperturbed reactor. If in equation 1.34 $\underline{\phi}$ were the flux after the reactor had been perturbed the equation would be exactly right and the result would be that of exact perturbation theory. It would not be nearly so useful, however, because equation 1.34 allows us to estimate the effect of any small perturbation using the results of just two multigroup calculations, one to find $\underline{\phi}$ from equations 1.30, the other to find $\underline{\phi}^*$ from equations 1.33.

1.2.6 Computations

The first step in calculating the properties of a reactor is to calculate the group fluxes and importances by solving equations 1.30 and 1.33. This has to be done on a large computer.

The information required for the analysis is the specification of the reactor and the nuclear cross-sections. The latter consist of data from thousands of measurements, so many that they can be manipulated only by computer. They are stored in magnetic data files which can be updated periodically as new experimental information becomes available.

Data are retrieved from the files by the calculation codes. The first step is usually a fundamental mode calculation in about 1000 fine groups for each region of the reactor under consideration. The fine group fluxes are then used to determine the group constants for a spatial calculation, usually in 30 or 40 groups. This solves equations 1.30 (i.e. equations 1.3) by an iterative method such as that described by Greenspan (1968). Apart from the "inner" iterations to find the flux distributions, however, there has to be an "outer" iteration to determine the eigenvalue, for equations 1.30 have no solution for $\underline{\phi}$ until the correct value of k has been found.

This outer iteration can be done in either of two ways. The composition and dimension of the reactor may be kept constant while the value of k is altered until the equations are solved. This is equivalent to finding the reactivity of a reactor which may not be exactly critical. Alternatively the composition (for example the concentration of plutonium in the core) or the dimensions (say the radius of the core) may be altered to make $k = 1$.

A three-dimensional reactor can often be represented by a two-dimensional model, usually in (r, z) cylindrical polar co-ordinates as most reactors are approximately cylindrical. Design calculations are usually done in this way, and the only important aspects of design which require the use of three-dimensional calculations are the determination of the effects of control rods and, especially for experimental reactors, of isolated abnormal fuel elements.

The requirements for calculations in support of an operating reactor, as opposed to one being designed, are rather different. It is necessary to predict, for example, the power generated in each fuel subassembly in advance of operation, so that it can be positioned in the core to achieve the correct burn-up and power during the next period of operation. A full three-dimensional representation of the reactor may be needed because account has to be taken of the slightly different composition of each subassembly, but this can often be offset by using fewer energy groups. Once suitable cross-sections have been determined for 5 or 10 very broad groups covering the entire energy range they are likely to be adequate for a range of calculations covering small variations in composition because the spectrum will not change very much. A typical approach to operational calculations is given by Wardleworth (1974).

1.2.7 Accuracy and Experimental Checks

Multigroup diffusion calculations can normally be expected to predict the reactivity of a system within about 0.5% and the distribution of power within about 4% throughout most of the core and within 6% close to control rods or the edge of the core. The errors are due partly to uncertainties in the basic cross-section data and partly to the approximations inherent in diffusion theory. The latter explain the larger errors near interfaces between different regions.

It is possible to check calculations by means of measurements in a reactor. In the past such experimental checks were very valuable but they have become less important as the reliability of calculation methods has been confirmed.

The most useful experimental techniques are to measure reaction rates or reactivity changes due to perturbations. Such measurements can be made on power reactors but it is often inconvenient to do so partly because it is difficult to obtain access to the core, partly because the temperature and neutron flux in the core may be too high for the measuring instruments, and partly because performance of reactor physics experiments conflicts with use of the reactor for the purpose for which it was built.

For this reason most, but not all, experimental checks on reactor physics calculations have come from zero power reactors. The most productive have been demountable facilities in which a reactor of almost any required composition can be assembled from samples of the various materials - structure, fuel, and even sodium coolant - which would be present in a power reactor. It can be made critical and operated at a power of typically a few watts so that no cooling is needed. Reaction rate measurements can be made with fission chambers or by activation methods, neutron spectra can be measured by proportional counters or time-of-flight techniques, and perturbation measurements can be made by inserting samples into the core and determining the resulting reactivity changes by moving a calibrated control rod to keep the reactor critical. The use of zero-power experimental reactors is described in more detail by Broomfield (1969).

1.3 FLUX DISTRIBUTION

1.3.1 Spectrum

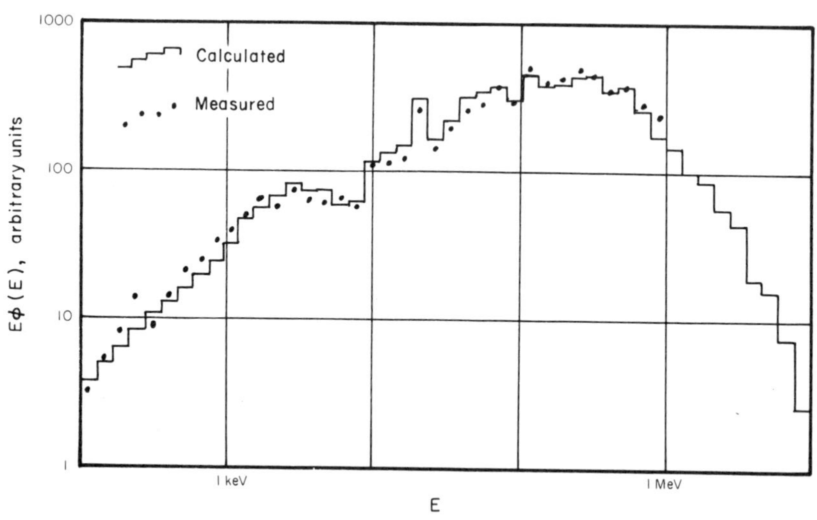

Fig. 1.5. Experimental and calculated neutron energy spectrum (Broomfield, 1969).

Figure 1.5 shows the relationship between calculated and measured spectra for a zero-energy experimental assembly similar in composition and size to a power reactor except that it contained carbon instead of oxygen and sodium. The spectrum was measured by a time-of-flight method, and is compared with the result of a 46-group fundamental mode calculation. The influence of the resonances in the iron cross-section on the flux can be seen clearly: where there is a resonance the measured flux is depressed, and if the resonance is wide enough compared with the group the calculated flux is also depressed. This is particularly noticeable for the iron resonance at 30 keV. The spectrum for a real power reactor would be very similar, the main difference being that there would be a depression due to the prominent sodium resonance at 3 keV.

TABLE 1.1 Dimensions and Compositions of Typical 600 MW (Heat) Reactor - (Equilibrium Core)

Radius	m	Height	m
Core Inner Zone	0.47	Core	0.92
Core Outer Zone	0.73	Upper Axial Breeder	0.49
Radial Breeder	0.91	Lower Axial Breeder	0.22
Reflector	1.21		

Compositions (Control Rods inserted into top 17% of Core)

	Fuel	Coolant	Structure	Absorber
Core	.286	.504	.207	.003
Upper Axial Breeder	.160	.549	.269	.022
Lower Axial Breeder	.265	.466	.269	0
Radial Breeder	.425	.302	.274	0
Reflector	0	.203	.797	0

Peak Burnup (%)

Core	3.75
Breeder	1.00

Fuel Composition

	Density	Enrichment	U Composition		Pu Composition			
	(kg m^{-3})	(Pu/(Pu+U))	^{235}U	^{238}U	^{239}Pu	^{240}Pu	^{241}Pu	^{242}Pu
Core Inner Zone	8644	.217	.003	.997	.765	.194	.036	.005
Core Outer Zone	8644	.279	.003	.997	.768	.192	.035	.005
Inner Breeder	9969	.037	.003	.997	.940	.057	.003	0
Outer Breeder	9969	.040	.003	.997	.918	.075	.007	0

Table 1.1 gives the compositions and dimensions of a typical 600 MW (heat) oxide-fuelled power fast reactor. The fuel composition is assumed to have reached equilibrium with a feed of plutonium from a thermal reactor (see section 1.4.5). The effect of different compositions is discussed below. The flux and importance at the centre of the reactor from a 16 group three-dimensional diffusion theory calculation are shown in Fig. 1.6. The actual spectrum would have a fine structure similar to that in Fig. 1.5.

Figure 1.7 shows the spectrum of neutrons as they are born in fission of ^{239}Pu. The fission spectrum for other isotopes is very similar. Comparison with Fig. 1.6 shows how the energy of the neutrons is reduced by scattering. At high energies, above about 0.5 MeV, inelastic scattering in ^{238}U and to a lesser extent in ^{56}Fe and ^{23}Na are very important. Excitation of the lowest energy level of the ^{238}U nucleus reduces the energy of the neutron by 45 keV, and the corresponding values

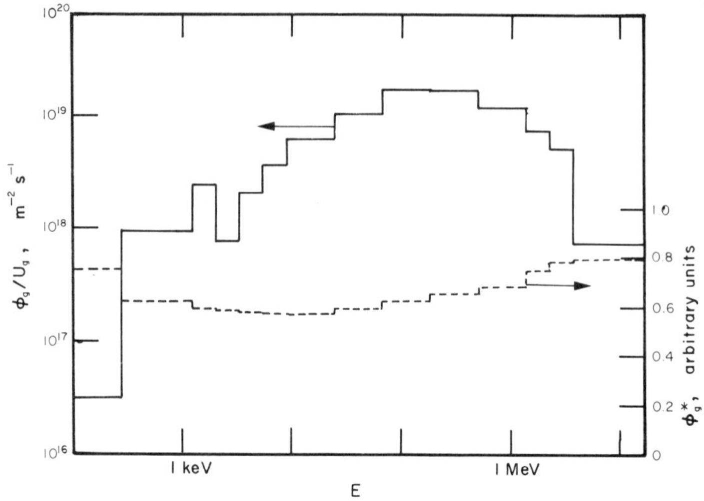

Fig. 1.6. Flux and importance at the centre of the reactor of Table 1.1.

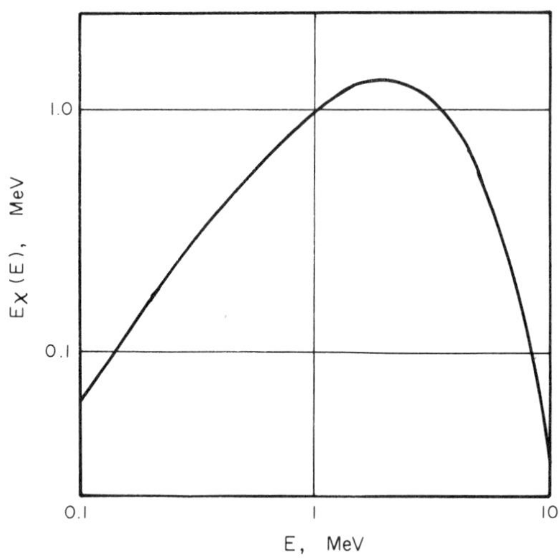

Fig. 1.7. Fission spectrum for ^{239}Pu.

for ^{56}Fe and ^{23}Na are 845 keV and 439 keV respectively, so the effect of inelastic scattering above 0.5 MeV is very marked.

At lower energies, elastic scattering reduces the energy of the neutrons, and the lightest nuclei such as ^{23}Na and ^{16}O are the most important moderators. More collisions are needed, however, to reduce the energy, and the chance that the neutron will be absorbed or will leak out of the core is large, so that the flux declines steadily with decreasing energy. Some of the neutrons are captured in ^{23}Na or ^{56}Fe but most are captured in ^{238}U or ^{240}Pu. Most of those leaking from the core are captured in ^{238}U in the breeder. Some leak from the breeder and are lost in the shield. This loss can be reduced by making the breeder thicker but a practical limit is set by the cost.

The importance, also shown in Fig. 1.6, is higher at high and low energies than intermediate. Above 1 MeV ϕ^* is high because of the possibility of fission in ^{238}U. At energies below 10 keV ϕ^* rises with decreasing energy because the fission cross-section in ^{239}Pu rises more rapidly than the capture cross-section in ^{238}U. The lower its energy the more likely a neutron is to cause fission and therefore contribute to the reactivity. It should be remembered that a neutron captured in ^{238}U is lost as far as maintaining the chain reaction is concerned even though it causes the generation of a new fissile nucleus.

1.3.2 Effects of Changes in Design

TABLE 1.2 Reactor Specification for Spectrum and Neutron Balance Calculations (Yiftah, 1961)

Compositions		
Plutonium	^{239}Pu	0.747
	^{240}Pu	0.102
	^{241}Pu	0.124
	^{242}Pu	0.027
Core Fuel	^{238}U:Pu adjusted to achieve criticality	
Core	Fuel	0.25
	Structure	0.25
	Coolant	0.50
Breeder	Fuel	0.60
	Structure	0.20
	Coolant	0.20
Reflector	Steel	1.00
Densities (kg m^{-3})		
Fuel	Oxide	8400
	Carbide	11390
Structure	Stainless Steel	7900
Coolant	Sodium	840
Dimensions (m)		
Core Radius	(1.5 m^3)	0.71
	(2.5 m^3)	0.84
Breeder Thickness		0.45
Reflector Thickness		0.30

TABLE 1.3 Effect of Fuel Material and Core Size on Neutron Balance (Yiftah, 1961)

Core Volume (m^3)	1.5		2.5		1.5	
Fuel Material	Oxide		Oxide		Carbide	
Critical Enrichment	0.212		0.182		0.158	
Critical Mass (kg total Pu)	590		845		645	
Neutron Balance	Core	Breeder	Core	Breeder	Core	Breeder
^{239}Pu fission	.244	-	.243	-	.236	-
capture	.0616	-	.0650	-	.0569	-
^{240}Pu fission	.00615	-	.00560	-	.00621	-
capture	.00816	-	.00851	-	.00758	-
^{241}Pu fission	.0558	-	.0564	-	.0530	-
capture	.00679	-	.00715	-	.00628	-
^{242}Pu fission	.00156	-	.00142	-	.00162	-
capture	.00212	-	.00221	-	.00202	-
^{238}U fission	.0296	.0235	.0320	.0184	.0413	.0208
capture	.168	.406	.212	.344	.225	.355
Coolant capture	.00772	.000608	.0102	.000560	.00628	.000509
Structure capture	.0253	.00676	.0297	.00565	.0221	.00598
Leakage from breeder	.0104		.00812		.00974	

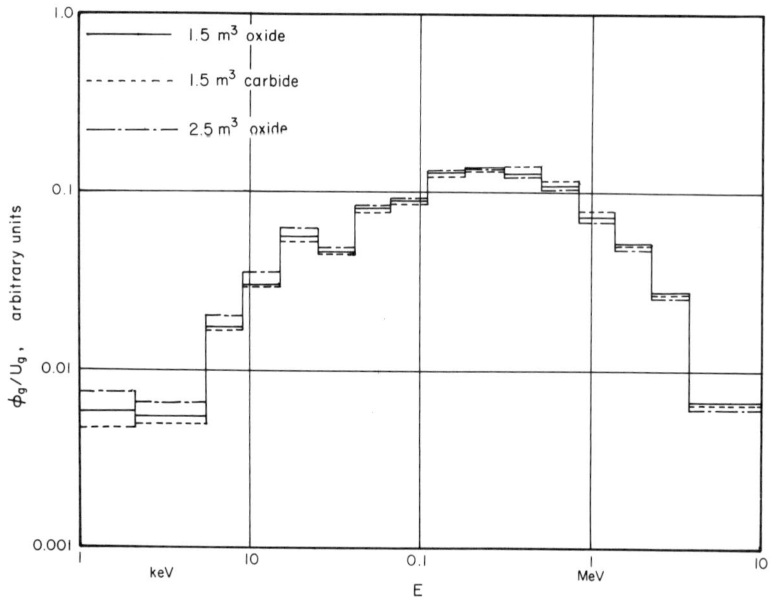

Fig. 1.8. Effect of fuel material and core volume on neutron energy spectrum.

Some effects of changes in core size and composition have been calculated by Yiftah (1961). Table 1.2 gives details of the composition of the spherical cores, surrounded by breeder and reflector, to which his calculations refer, and some of the results are given in Table 1.3 and Fig. 1.8. Table 1.3 gives neutron balance sheets for various reactors and shows the fates of the neutrons generated in fission. The numbers given are the total reaction rate for the reaction in question, integrated over all energies and over the whole core or breeder, and divided by the total rate of generation of fission neutrons in the core. The numbers add up to a little more than one because the small number of fissions in the breeder contribute a few extra neutrons.

The effect of changing the size of the reactor is shown by comparing columns 1 and 2 of Table 1.3. As the core becomes larger fewer neutrons leak out. These extra neutrons are absorbed by increasing the amount of fertile material in the core and decreasing the amount of fissile material, so that the proportion of fissile material in the fuel, often called the "enrichment", is decreased while the total mass of fissile material, the "critical mass", goes up. The total amount of breeding decreases slightly mainly because of the slight reduction of fission in the breeder, but the effect is slight. More important is the fact that the amount of breeding in the core increases.

The effect of replacing oxide by carbide fuel is shown by comparing columns 1 and 3 of Table 1.3. Fewer moderating atoms are present in a UC/PuC mixture than in UO_2/PuO_2 and even though a carbon nucleus is lighter than oxygen there is less moderation and the mean neutron energy is higher. This is indicated by the fact that the peak in the spectra in Fig. 1.8 is at a higher energy for carbide than for oxide. This is often called a "harder" spectrum.

At the peak of the spectrum neutron importance increases with energy (ϕ^* is higher), so that the harder spectrum resulting from the change from oxide to carbide fuel allows fissile material to be replaced by fertile and the enrichment is lower. There are more captures in the fertile material in the core, and although there are fewer in the breeder the gain outweighs the loss and there is a net increase in breeding. The density of carbide is greater than that of oxide so that the quantity of both fissile and fertile material in the core is increased and, in spite of the lower enrichment, the critical mass is higher. Similar, but greater, changes result if oxide is replaced by metal.

1.3.3 Power Distribution and Enrichment Zones

If the core had the same composition throughout the power density would be nearly proportional to the flux and would be distributed across the core as shown in Fig. 1.9. This would be most undesirable from a thermodynamic point of view because if the coolant flowed at the same rate through all parts of the core it would emerge at different temperatures. When cold coolant from the periphery of the core mixed with hot coolant from the centre there would be a gain of entropy and a consequent loss of work output, and there would be large temperature fluctuations in the mixing region which might damage the structure (see section 3.3.7). Alternatively if the flow in the outer part of the core were restricted to equalise the outlet temperatures pumping work would be wasted as the coolant flowed through the restrictions.

For this reason the core is normally made in two or more radial zones, with higher fuel enrichment in the outer zones. The effect of this is to increase the power density in the outer region as shown in Fig. 1.10. In the reactor shown there are two zones of roughly equal volume, with enrichments of 22% and 28%. If the core were uniform the enrichment would be about 24%. The peak power densities at

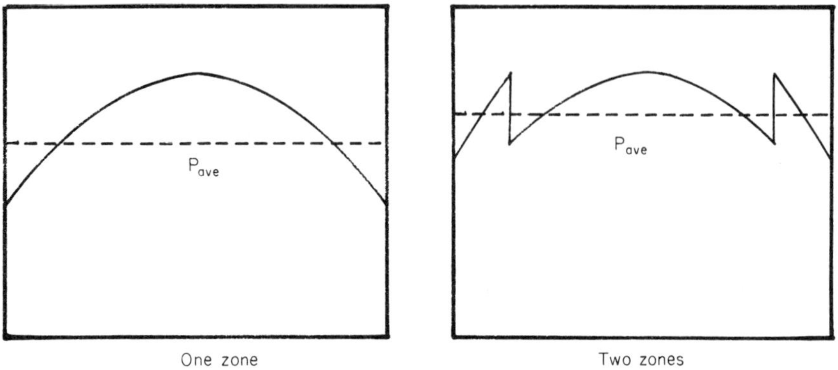

Fig. 1.9. Effect of enrichment zones on radial distribution of power.

the centre of the core and the inside of the outer zone are roughly the same and the radial power peaking factor, P_{max}/P_{ave}, where P_{max} is the power generated in the most highly rated channel and P_{ave} is the average power per channel, is reduced from about 1.35 in a single-zone core to 1.21 in the two-zone core. The axial distribution of power with control rods partially inserted is shown in Fig. 1.11.

An important point to notice in both Figs. 1.10 and 1.11 is the change in the power density at the inside of the breeder with time. As fissile material is generated in the breeder, predominantly at the inside, it in turn undergoes fission and generates power. Over the life of the inner fuel elements of the radial breeder (which would stay in the reactor much longer than the core fuel) the power density rises considerably. In the reactor of Table 1.1, at the start of its life the power density at the centre of an inner radial breeder fuel element is 60 MW m^{-3} when the reactor power is 600 MW (heat), but this rises to 220 MW m^{-3} after 1.6 years at power.

1.4 BREEDING

1.4.1 Breeding Processes

As we have seen breeding is basically the conversion of ^{238}U into ^{239}Pu by neutron capture. This simple picture is complicated, however, by several other processes which occur. ^{239}Pu is the beginning, not the end, of a chain. Neutrons are captured in it to convert it first to ^{240}Pu. This is a fertile isotope with properties rather like those of ^{238}U: it is fissile to neutrons above 1 MeV, and at all energies has a substantial capture cross-section. It is radioactive, decaying by α-emission (as do all the isotopes of uranium and plutonium) with a half-life of 6540 years.

Neutron capture in ^{240}Pu produces ^{241}Pu. This is a fissile isotope with properties similar to those of ^{239}Pu except that its fission cross-section, especially below

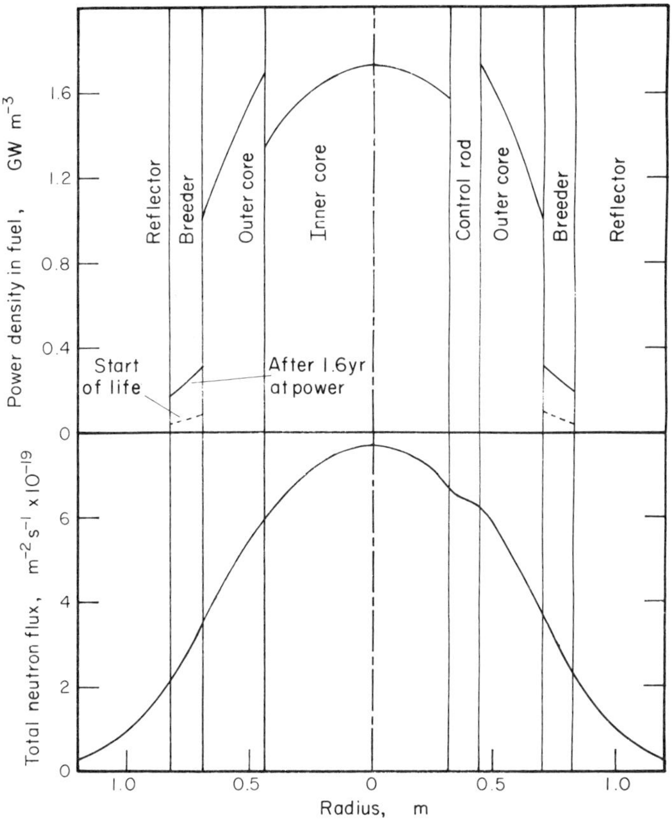

Fig. 1.10. Radial distribution of total flux and power density in fuel for reactor of Table 1.1.

0.1 MeV, is higher. It also captures neutrons producing ^{242}Pu which again is fertile with a fission threshold at 1 MeV. This is the last of the chain of plutonium isotopes produced: neutron capture in ^{242}Pu produces ^{243}Pu but this is β-active with a half-life of 5 hours, decaying into ^{243}Am. Still higher isotopes are formed by further neutron capture reactions but the quantities produced are small and with two exceptions they can be neglected as far as reactor operation is concerned.

The exceptions are two isotopes of curium, ^{242}Cm and ^{244}Cm. The routes by which they are formed from ^{241}Pu can be represented diagrammatically as follows, where the horizontal arrows represent (n,γ) reactions while the vertical ones represent β^- or β^+ decay as shown.

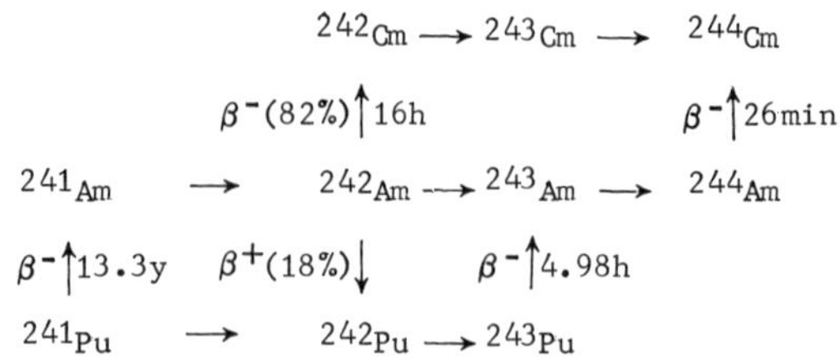

All these nuclides decay by spontaneous fission but ^{242}Cm and ^{244}Cm have much shorter half-lives than the rest, 7.2×10^6 y and 1.4×10^7 y respectively. The neutrons generated are insignificant in normal operation of the reactor but not when it is shut down. As the quantity of ^{241}Pu changes with burnup, so do the quantities of the curium isotopes and hence the neutron source strength. As a result the relationship between shut-down reactivity and subcritical power level varies both with burnup and also with time while the reactor is shut down due to decay of ^{242}Am and ^{244}Am.

There are further complications. The fertile ^{238}U in the reactor is not pure but is usually depleted uranium from a fuel enrichment plant. No enrichment process removes all the ^{235}U from natural uranium and the depleted uranium in a fast reactor may well contain 0.25% of ^{235}U. Neutrons are captured in this to form ^{236}U which, being even-numbered, is another fertile isotope. There may also be some ^{234}U present which is converted to ^{235}U.

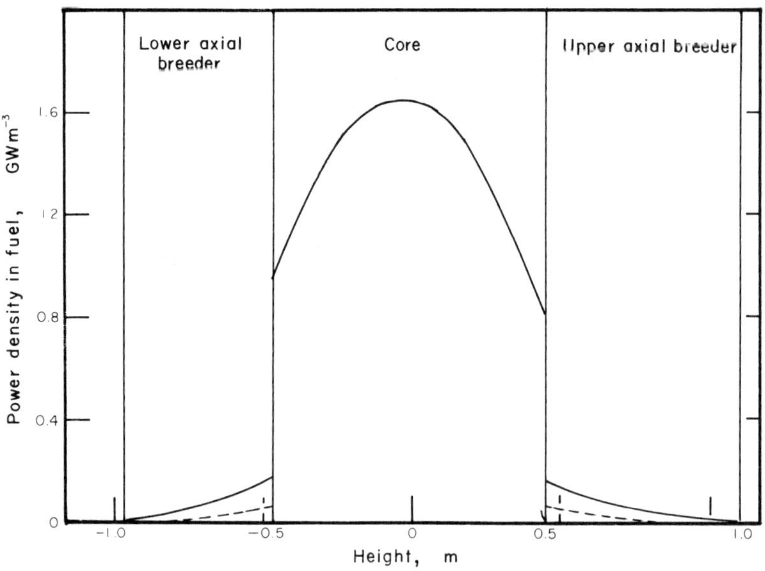

Fig. 1.11. Axial distribution of power density in fuel for the reactor of Table 1.1.

1.4.2 Breeding Ratio

For most purposes it is adequate to neglect these complications and concentrate on the five isotopes of greatest importance, ^{238}U, ^{239}Pu, ^{240}Pu, ^{241} Pu and ^{242}Pu. In the introduction we gave a simple definition of the breeding ratio B as being the ratio of the number of fissile nuclei produced to the number consumed. This can be written as

$$B = \frac{C_8 + C_0}{A_9 + A_1} \tag{1.35}$$

where C represents capture rate and A absorption rate (absorption meaning either capture or fission), and the subscripts 8, 9, 0 and 1 represent the isotopes ^{238}U, ^{239}Pu, etc. Each of the reaction rates is the total over the whole reactor and for all neutron energies so that in terms of the group fluxes we have for example

$$C_8 = \int_R \sum_g \Sigma^8_{cg} \phi_g \, dv, \tag{1.36}$$

where Σ^8_{cg} is the capture cross-section for ^{238}U in group g. The suffix R denotes that the integral runs over the whole reactor including the breeder.

B is widely used as a measure of the effectiveness of a breeder reactor, but it suffers from two disadvantages. The first is the minor one that some confusion arises if there is a significant amount of ^{235}U in the core. The major disadvantage, however, is the fact that it does not take account of the differences between the isotopes. It counts an atom of ^{241}Pu as being equal to an atom of ^{239}Pu, whereas in fact the fission cross-section of ^{241}Pu is higher than that of ^{239}Pu and its capture cross-section is lower. It is therefore more valuable as a reactor fuel and if, for example, we had two breeder reactors with the same B, but one of which produced ^{241}Pu while the other produced ^{239}Pu, the first would produce enough fuel to start another reactor more quickly than the second.

1.4.3 Breeding Gain

A better measure of breeding is the "Breeding Gain" introduced by Baker (1963). This is based on an assessment of the various isotopic changes using perturbation theory in a simplified form for convenience.

If there is only one neutron energy group we can see from equations 1.30 and 1.33 that neutron flux and importance, ϕ and ϕ^*, are proportional. If equation 1.29 is then rewritten for one group we have

$$\delta\rho \;\alpha \int_R \left\{ \delta(\bar{\nu}\Sigma_f - \Sigma_f - \Sigma_c) \, \phi^2 + \delta D \, (\nabla\phi)^2 \right\} dV, \tag{1.37}$$

because $\Sigma_r \equiv \Sigma_f + \Sigma_c$, and the scattering terms are no longer significant. The macroscopic cross-sections are given by expressions like

$$\Sigma_f = \sum_i N_i \, \overline{\sigma_{fi}}, \tag{1.38}$$

where the subscript i represents the nuclides of which the reactor is composed,

N_i is the number of nuclei of i per unit volume, and $\overline{\sigma_{fi}}$ is the average microscopic fission cross-section of i. The average is over the entire neutron energy range, and weighted with $\phi\phi^*$; i.e.

$$\overline{\sigma_{fi}} = \int_0^\infty \sigma_{fi}(E)\,\phi(E)\,\phi^*(E)\,dE \Big/ \int_0^\infty \phi(E)\,\phi^*(E)\,dE, \qquad 1.39$$

where $\phi(E)$ and $\phi^*(E)$ are the fundamental mode flux and importance.

It follows from equation 1.37 that, if the effect of scattering is neglected, the reactivity increase when an atom of i is created in the reactor is proportional to x_i, where

$$x_i = \overline{\nu\sigma_{fi}} - \overline{\sigma_{fi}} - \overline{\sigma_{ci}}. \qquad 1.40$$

x_i thus measures the usefulness of an atom of i for building a new reactor. The "worth" of i, w_i, is defined by

$$w_i = (x_i - x_8)/(x_9 - x_8), \qquad 1.41$$

where subscripts 8 and 9 refer to ^{238}U and ^{239}Pu respectively. w_i is thus the reactivity effect of an atom of i measured on a scale on which the worth of an atom of ^{238}U is 0 and of ^{239}Pu is 1.

The "Breeding Gain", G, is then defined by

$$G = \sum_i w_i\,(C_{i-1} - A_i) \Big/ \sum_i F_i. \qquad 1.42$$

C_{i-1} and A_i are the rates of creation and destruction of atoms of i, and the denominator of the expression for G is the total fission rate. The sums are taken over all the fuel isotopes, both fissile and fertile, present in the reactor. G is thus proportional to the increase in reactivity of the fuel per fission, assuming that the fuel is continually being removed and the excess incorporated uniformly throughout a new reactor. Table 1.4 gives values of B and G for the reactors described in Table 1.2 (Section 1.3.2). It will be seen that $G \simeq B - 1$, as is to be expected from the way the terms are defined, but there is no simple algebraic way of showing that it must be so.

TABLE 1.4 Effect of Fuel Material and Core Size on Breeding Gain and Breeding Ratio

Core Volume (m^3)		1.5	2.5	1.5
Fuel Material		Oxide	Oxide	Carbide
	w_8	0	0	0
	w_9	1	1	1
"Worth" of Nuclides	w_0	0.151	0.131	0.163
	w_1	1.507	1.541	1.485
	w_2	0.140	0.121	0.153
Breeding Gain	G	0.54	0.48	0.61
Breeding Ratio	B	1.57	1.52	1.67
Internal Breeding Gain	G_I	-0.59	-0.49	-0.38
Internal Breeding Ratio	B_I	0.48	0.59	0.66

The advantage of measuring breeding in terms of G rather than B stem from the fact that the denominator of G is proportional to the reactor power and the numerator to the rate at which new critical masses can be assembled. It should be noted that G can be used to take account of any sort of fuel, containing ^{233}U and ^{235}U for example, and any of the breeding reactions.

On the other hand it has an important limitation in that it assumes that the excess fuel is to be used in an identical reactor. If the fuel were to be used to make a different reactor (of different size, for example, or with carbide rather than oxide fuel), G would not be the appropriate measure of its worth, and although G is nearly the same for all large fast reactors it is very different for a thermal reactor. The dependence on the nature of the reactor, which is not apparent in equation 1.43, lies in equation 1.40. In a different reactor ϕ and ϕ^* will be different and different average cross-sections will result. The effect of this is shown by the different values of w_i for the different reactors in Table 1.4.

1.4.4 Internal Breeding

G as defined in equation 1.43 measures the gain in reactivity assuming the fuel is to be used in a new reactor. Breeding also has an effect on the reactivity of the reactor in which it takes place, but this is not related at all closely to the value of G. Table 1.3 shows, for example, that much of the new fissile material is generated in the breeder where its effect on reactivity is very small as both ϕ and ϕ^* are low (Fig. 1.10). The fissile material generated within the core does have a significant effect on reactivity, however, which is very important. This can be characterised by the "internal breeding gain", G_I, which is defined in exactly the same way as G except that the capture and absorption rates C_{i-1} and A_i in the numerator of equation 1.43 are integrated over the core of the reactor, not over the whole reactor including the breeder as indicated in equation 1.36. Values of G_I are also given in Table 1.4. G_I is negative for most reactors, implying that there is a nett loss of fissile material from the core, but it can be positive for metal-fuelled or gas-cooled fast reactors.

G_I does not measure the reactivity change directly. In the same way as G it represents the effect on reactivity which would result if the changes in composition of the fuel in the core were distributed uniformly throughout the core, whereas in fact the changes are greatest at the centre of the core where the flux is highest. In spite of this G_I is a useful guide to reactivity changes and when G_I is greater (i.e. less negative) the loss of reactivity with burnup is less.

It is important that the rate of change of reactivity with burnup should be as small as possible. The fuel is changed in batches when the reactor is shut down, and it is desirable to make the interval between changes as long as possible to minimise interruption to operation. Between changes the reactivity of the fuel declines and has to be compensated for by withdrawing control rods containing neutron absorbers from the core. When the fuel is fresh, therefore, neutrons which could have been used for breeding are absorbed in the control rods. The higher the internal breeding the smaller the reactivity loss with burnup, the smaller the quantity of neutron absorbing material in the core and the better the overall breeding or, alternatively, the longer the period between fuel changes.

1.4.5 Fuel Composition

The fuel for a new fast reactor may not come from another fast reactor and indeed in the early years of the exploitation of breeder reactors is unlikely to do so. The first fast reactors will presumably be fuelled with plutonium from a thermal reactor or possibly with enriched uranium from an enrichment plant. Only later will it be possible to fuel new fast reactors with plutonium from the breeders of existing fast reactors.

The isotopic composition of the plutonium from the various sources is different, and has an effect on the operation of the reactor. Plutonium from a fast reactor breeder consists mainly of ^{239}Pu, while plutonium from a thermal reactor has greater concentrations of the higher isotopes. Table 1.5 shows the effect of irradiation on the composition of plutonium from a thermal reactor.

TABLE 1.5 Effect of Thermal Reactor Fuel Burnup on Plutonium Composition

Source	Plutonium Composition (%)			
	239	240	241	242
Magnox fuel, irradiated to 3000 MWd/tonne	.800	.169	.027	.004
AGR fuel, irradiated to 20000 MWd/tonne	.56	.20	.15	.09

The fuel composition changes as the reactor is operated until it is refuelled. Then a fraction, typically a third, of the fuel elements are removed and replaced with fresh ones. Those that are removed are reprocessed to remove the accumulated fission products, to replace the lost fissile material with plutonium from the feed source, and to make up the lost ^{238}U with depleted uranium. They are then available to be returned to the reactor at a subsequent refuelling.

TABLE 1.6 Effect of Fuel Feed Source on Equilibrium Composition, Breeding and Critical Mass (Okrent, 1961) - (1.5 m^3 core, oxide fuel)

		Feed of Pure ^{239}Pu		Feed of Light-water Reactor Pu	
		Initial	Equilibrium	Initial	Equilibrium
Fuel Composition	^{238}U	.767	.696	.754	.678
	^{239}Pu	.233	.188	.183	.161
	^{240}Pu	0	.095	.025	.112
	^{241}Pu	0	.015	.030	.030
	^{242}Pu	0	.006	.007	.018
Breeding Ratio		1.47	1.67	1.57	1.80
Critical Mass (kg of ^{239}Pu + ^{241}Pu)		562	496	514	468

Figure 1.12 shows the variation of mean fuel composition through several refuelling cycles for reactors fed with pure ^{239}Pu and with thermal reactor plutonium, as calculated by Okrent (1961). From its initial composition the fuel approaches an equilibrium composition after 10 cycles or so, about which it oscillates. The equilibrium composition depends on the composition (and hence on the source) of the feed. Table 1.6 gives the initial and equilibrium fuel compositions of reactors fed with pure ^{239}Pu and plutonium from a light-water reactor.

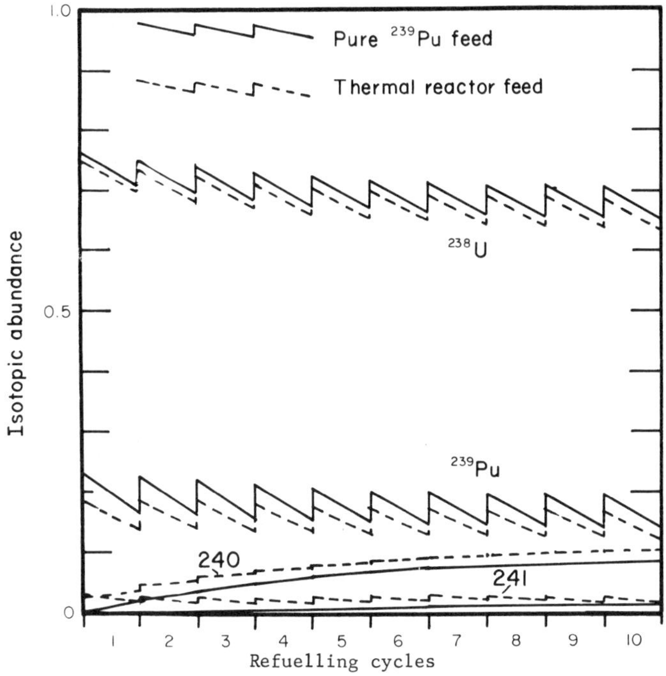

Fig. 1.12. Effect of burnup on fuel composition.

Breeding and critical mass also depend on the composition of the plutonium feed. Breeding is higher if the feed is of thermal reactor plutonium containing higher isotopes than if it is pure ^{239}Pu, and the critical mass (i.e. the mass of ^{239}Pu and ^{241}Pu, not the total mass of plutonium) is smaller. This is because ^{241}Pu has a higher fission cross-section than ^{239}Pu, and ^{240}Pu has a higher fission cross-section and a lower fission threshold than ^{238}U. These effects are also shown in Table 1.6.

1.5 CONTROL RODS

1.5.1 Materials

The reactor is controlled and shut down by moving neutron absorbing material into or out of the core. Other methods of control such as moving fuel or parts of a neutron reflector around the core were used in early experimental reactors but are not as convenient in a large power reactor.

The absorbing material is usually boron, possibly enriched in ^{10}B, in the form of the carbide B_4C. Alternatives are metallic tantalum, or oxides of europium or gadolinium. ^{10}B has a very high capture cross-section which varies as $E^{-\frac{1}{2}}$ up to 0.1 MeV and it captures neutrons predominantly at the low-energy end of the spectrum.

A disadvantage of boron is that it captures neutrons by an (n, α) reaction:

$$^{10}B + n \rightarrow {}^{7}Li + {}^{4}He,$$

so that while it is in use in an operating reactor helium atoms accumulate within the crystals of the boron carbide. These tend to form little bubbles of gas which disrupt the structure of the crystals and damage the material. This, together with loss of the ^{10}B, limits the life of a rod used for controlling an operating reactor. Shut-off rods, used only to shut the reactor down and hold it well sub-critical during refuelling, are not subject to this limitation.

1.5.2 Reactivity Worth

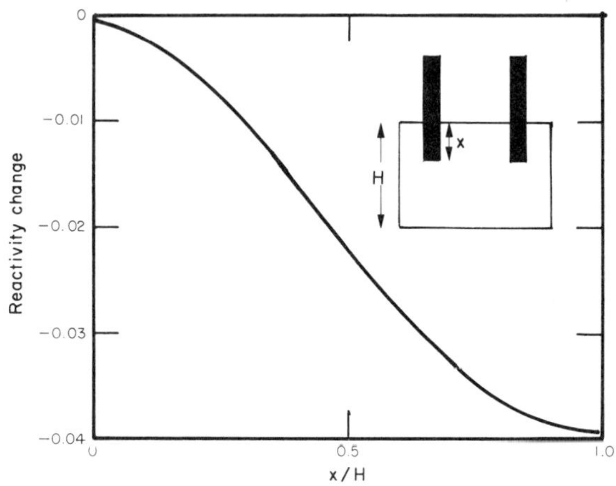

Fig. 1.13. Reactivity change due to insertion of control rods.

If a control rod is inserted a distance x into the core, as shown in Fig. 1.13, the change in reactivity (neglecting any effects other than neutron capture) may be estimated to be proportional to the integral of $\phi\phi^*$ over the length of the rod times the difference between the capture cross-section of the material making up the rod and that of the material (presumably coolant) it displaces - i.e.

$$\Delta\rho \propto \int_0^x \sum_g \phi_g \, \phi^*_g \, (\Sigma_{cg}^{rod} - \Sigma_{cg}^{coolant}) \, dx. \qquad 1.44$$

The result is an S-shaped curve of reactivity against position as shown in Fig. 1.13. The reactivity change on inserting a control rod fully into the core is usually called the "worth" of the rod.

This first-order perturbation theory estimate is not an accurate indication of the worth because the distributions of ϕ and ϕ^* are altered by the presence of the rod. Figure 1.14 shows the axial variation of ϕ (actually the total flux summed over all

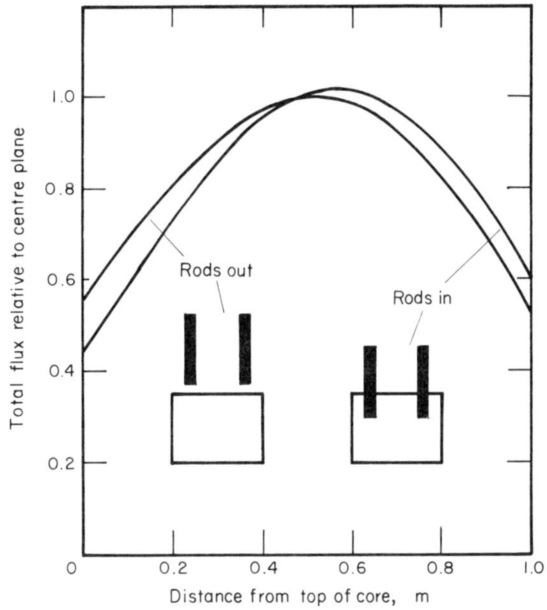

Fig. 1.14. Effect of partially-inserted control rods on flux distribution.

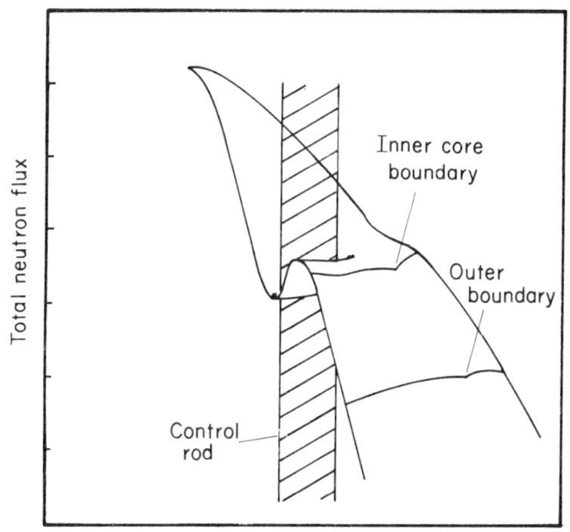

Fig. 1.15. Flux depression round a control rod.

groups) with control rods withdrawn and with all the control rods inserted a third of the way into the core. The presence of the rods in the top of the core pushes the flux towards the bottom. Figure 1.15 shows the flux distribution around a single partially-inserted rod and indicates how the flux is depressed in its vicinity.

The result is that the reactivity worth of one control rod depends on the position of all the others, and there are many control rods and shut-off rods in a large reactor. The worth of a rod will be lower if the neighbouring rods are inserted and higher if they are not.

It is normal, however, to move the control rods together to keep flux distortion across the core to a minimum. If this were not done the power density might be higher on one side of the core than the other causing greater non-uniformity of coolant outlet temperature and therefore more entropy gain due to mixing. There would also be variations of the fuel burnup rate.

TABLE 1.7 Reactivity Requirements and Control, Shut-off and Safety Rods

Source	Reactivity Increase
A Temperature (from shutdown condition to operation at low power)	-0.6%
B Power (from low to full power)	-1.0%
C Burnup (typically 1% between fuel changes)	-2.4%
D Shutdown Margin	-2.4%
Total	-6.4%

These requirements may be met as follows.

	Reactivity Worth
Control rods (covering reactivity changes during operation) A + B + C	-4.0%
Shut-off rods (to hold reactor sub-critical during refuelling) D	-2.4%
Safety rods (independent system capable of shutting reactor down at any time) A + B	-1.6%

The total worth of the control rods is typically some 4%, made up as indicated in Table 1.7. The "shut-down margin" is a safety allowance to make sure that the reactor remains sub-critical even if errors are made during refuelling and excessive fuel is added. The worth of individual rods is chosen so that the reactor can be shut down from any operating condition even if some of the rod mechanisms fail to ... also so that the reactivity increase if any one rod is accidentally with-... damage. These points are discussed in more detail in

1.6 REACTIVITY COEFFICIENTS

1.6.1 Effects of Temperature

Temperature affects reactivity in a number of ways. The temperature of the structure affects the dimensions of the reactor and sometimes the relative position of the various parts: the densities of all the materials depend on temperature (but the most important effects arise from changes in the density of the coolant); and the temperature of the fuel affects the resonance self-shielding in the fuel materials. The various effects are discussed in detail by Hummel (1970).

If the temperature changes are small the resulting changes in dimensions, density and self-shielding are also small in most cases and proportional to the temperature changes. First-order perturbation theory is valid and the resulting reactivity changes are, approximately at least, linear and independent. As a result it is useful to express temperature-induced reactivity changes as reactivity coefficients of the form $\partial\rho/\partial T_i$, where T_i is the temperature in question (the coolant inlet temperature, for example, or the mean fuel temperature), and $\partial\rho/\partial T_i$ may be taken to be constant and independent of all the T_i.

This approximation breaks down in some cases. It may not be true in normal operation if there is intermittent contact between bowing fuel elements, and it is certainly untrue in the extreme conditions which may be encountered in an accident - if the coolant boils, for example, or if the fuel temperature rises very high.

1.6.2 Structure Temperatures

The effect of the temperatures of the many different parts of the reactor structure on reactivity will depend on the detailed design of the reactor. The possible overall effects can be illustrated by the following examples, but the reality may be much more complex.

The radial dimensions of the reactor are determined by the temperature of the structure which supports it while the axial dimensions of the core may depend mainly on the temperature of the fuel cladding, so radial and axial dimensions may change independently. If the structural temperatures increase the mean "smeared" densities of the solid materials decrease, but the coolant mean density may remain the same or it may actually increase. If, for example, the temperatures of all the materials in the core are constant, while the temperature of the structure increases so that the core radius increases by 1%, the core volume will increase by 2%. The actual volume and mass of fuel and structural material in the core will remain the same, so their "smeared" densities will decrease by 2%. If the fuel and structure made up say 50% of the original core, however, they will form only 49% of the expanded core, so that the coolant volume fraction increases from 50 to 51% and its "smeared" density will increase by 2%.

Figure 1.16 shows the effects of 1% increases in the radial and axial dimensions of the whole reactor for a typical oxide-fuelled core as a function of the ratio of height to radius, H/R. The reactivity effect has two components - an increase due to the increased size and a decrease due to the decreased density. In all cases the density effect is greater and reactivity decreases with expansion, but if H/R is small the effect of increased height is small. In principle, in the limit of an infinite slab reactor ($R \rightarrow \infty$), uniform expansion in the axial direction has no effect on reactivity at all.

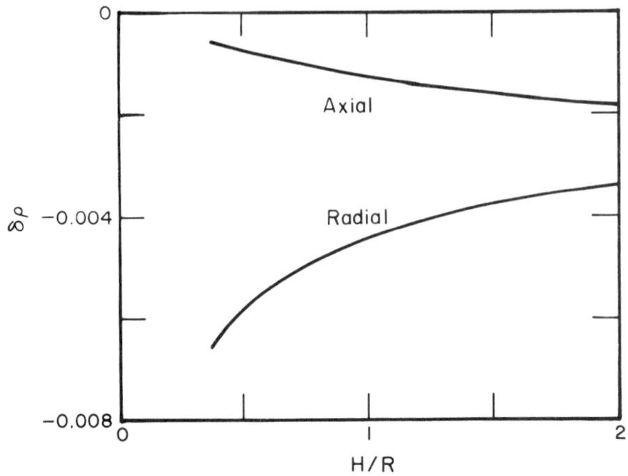

Fig. 1.16. Effect of core shape on reactivity change due to 1% axial and radial expansion.

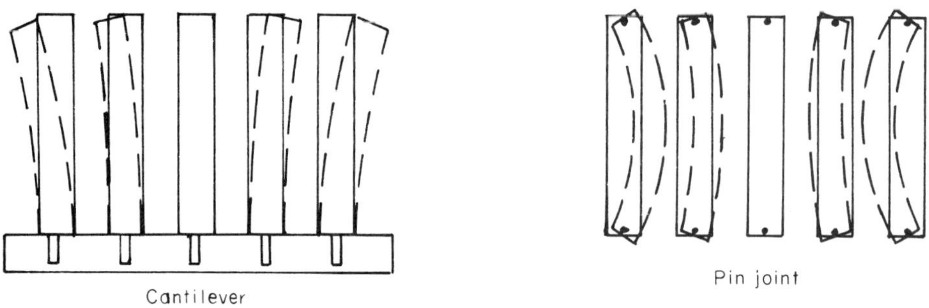

Fig. 1.17. Effect of fuel support on bowing.

1.6.3 Bowing

As the power density is higher at the centre of the core than at the edge there is a tendency for fuel elements in the outer part of the core to be hotter on the side facing the core centre. If they are straight when they are cold - i.e. when the reactor is shut down - they will tend to become curved when the reactor is at power, with the convex side towards the core centre.

The effect on reactivity depends on how they are supported. Figure 1.17 shows that if they are cantilevered at one end they move outwards as the power increases, reducing reactivity, but if they are pin-jointed at both ends they move inwards, adding reactivity and giving positive feedback.

In practice, however, the fuel elements cannot move freely very far before they touch each other, and the reactor may be designed so that they cannot move freely at all. For this reason the effect of bowing on reactivity is likely to be small, but there may be complicated non-linear effects as fuel elements or subassemblies distort to take up small clearances within the manufacturing tolerances.

1.6.4 Coolant Density

Equation 1.29 shows three effects of a change in density of a non-fissile isotope: a "moderating" effect due to the change in $\Sigma_{sg\to g'}$, a "capture" effect due to the change in Σ_{rg}, and a "scattering" effect due to the change in D_g. There is a fourth "self-shielding" effect which is shown in equation 1.23: if Σ_s is changed all the group cross-sections for the resonant isotopes, such as Σ_{cg} in equation 1.21, are altered.

As a result the effects of an increase in the coolant temperature are as follows. The reduction in the density of the sodium reduces Σ_{cg} giving an increase in reactivity and a positive capture contribution to the sodium temperature coefficient, but it is a small contribution because the sodium does not capture many neutrons in the first place (see Table 1.3).

The reduction in $\Sigma_{sg\to g'}$ has the effect of hardening the spectrum - that is of shifting the peak in ϕ shown in Fig. 1.16 to a slightly higher energy. The peak occurs at about 0.1 - 0.3 MeV, where ϕ^* increases with energy, so there is a gain in reactivity. Because sodium is an effective moderator the positive moderating contribution to the sodium temperature coefficient is large.

As Σ_s decreases Σ_{cg} decreases because ϕ_g in equation 1.21 increases more than J. There is a similar effect on Σ_f but, as in the case of a change in the fuel temperature, the capture effect from ^{238}U is greater than the fission effect from ^{239}Pu and ^{241}Pu, so the reactivity increases giving another small positive contribution to the sodium temperature coefficient.

The capture, moderating and self-shielding effects are contributions to the $\phi_g\,\phi^*_g$ term in equation 1.29, so they all depend on position in roughly the same way. They are greatest at the centre of the core and smaller at the edges. The effect of scattering is quite different, however. A decrease in the scattering cross-section increases the diffusion coefficient D_g and this results in a decrease in reactivity depending this time on $\nabla\phi_g.\nabla\phi^*_g$. It is therefore zero at the centre and reaches a maximum in the outer parts of the core.

Table 1.8 shows the effect of increasing the coolant temperature throughout the core of one particular reactor by the same amount. The resulting reactivity change is the small difference between two quantities, and in this case it is just positive giving an overall positive sodium temperature coefficient.

The effect of a local change in sodium temperature or of a local loss of sodium may, on the other hand, be positive or negative depending on where in the core it happens. Figure 1.18 shows the effect of loss of coolant from various points on the axis of the core. Near the centre moderating, capture and self-shielding dominate and the reactivity change is positive, but towards the edges leakage becomes more important and it is negative.

Complete loss of sodium from all or part of the core might be imagined to happen in a very serious accident in which the coolant boils. The reactivity increase due to complete loss of coolant from the core of the reactor of Table 1.8 is 4×10^{-5}, but the increase can be as great at 7×10^{-3} if coolant is lost just from the part of

TABLE 1.8 Components of the Isothermal Sodium Temperature Coefficient of Reactivity for a Small Fast Reactor Core

Reactor Specification

Core Composition (by volume)	Fuel	.34
	Coolant	.45
	Structure	.21
Fuel Enrichment (Pu/(Pu + U))	Inner Core	.19
	Outer Core	.29
Plutonium Composition	239	.800
	240	.169
	241	.027
	242	.004
Core Dimensions (m)	Height	0.8
	Diameter	1.34

Sodium Temperature Coefficient of Reactivity (K^{-1})

(Core only)

Moderation	5.22×10^{-6}
Capture	0.78×10^{-6}
Self-shielding	0.52×10^{-6}
Scattering	-6.51×10^{-6}
Total	0.01×10^{-6}

the core where the sodium coefficient is positive. In a large 2500 MW (heat) reactor the reactivity increases would be more positive, typically .017 for complete loss of coolant and .020 for loss from the central region.

1.6.5 Doppler Coefficient

As the fuel temperature increases the resonance self-shielding in the fuel isotopes decreases. The resulting change in reactivity is called the Doppler effect because it is due to the dependence of the effective energy of a neutron on the relative motion of the nucleus with which it interacts. The decrease in self-shielding results in an increase in all the cross-sections of the fuel isotopes at energies below about 20 keV.

Although all the cross-sections increase including scattering the only significant changes are in Σ_c for ^{238}U and Σ_f for ^{239}Pu and ^{241}Pu. In all cases the increase in capture in ^{238}U dominates and the Doppler effect is negative.

It is not independent of temperature however. The dependence of J (equation 1.22) on T is complex. If β/ψ_o, where ψ_o is the peak value of ψ, is large (corresponding to a resonance with a low peak at higher energy) then $J \propto T^{-1/2}$ and $dJ/dT \propto T^{-3/2}$, but if β/ψ_o is small $J \propto T$ and dJ/dT is constant. The Doppler effect thus varies in a different way with temperature for different resonances but the total coefficient, involving a weighted sum of dJ/dT for all the resonances, varies roughly as T^{-1}. As a result it is convenient to define a "Doppler constant", $- T\, d\rho/dT$.

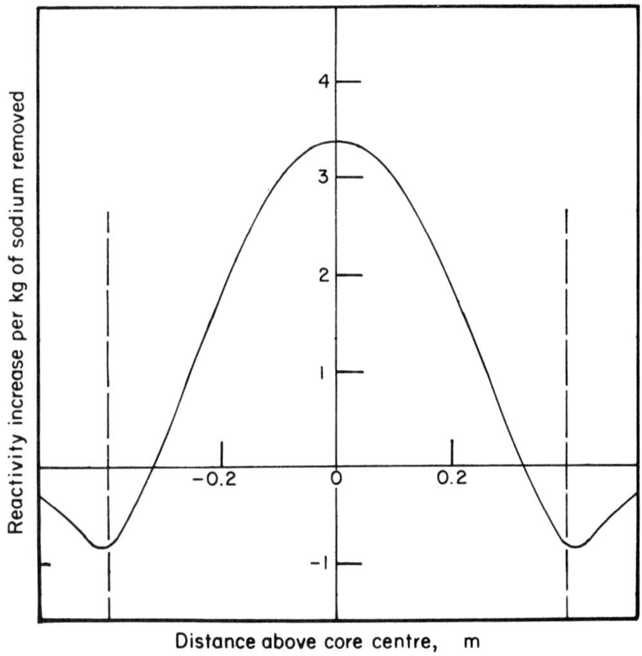

Fig. 1.18. Reactivity change on loss of coolant as a function of position.

Figure 1.19 shows the contribution to the Doppler reactivity effect from various isotopes in different neutron energy groups for a reactor of simple composition and low enrichment. The most important effects are around 1 keV at which energy the resonances are not resolved. The effect of the iron resonance at 1 keV can be seen. This is the only resonance in a non-fissile or fertile isotope which makes any significant contribution to the Doppler effect.

In an operating reactor the fuel is not all at the same temperature, and as the Doppler coefficient is a function of temperature there is a question of the correct average temperature to use. In a cylindrical fuel element with constant thermal conductivity the volume-weighted mean temperature $\overline{T}$ is half way between the central and surface temperatures. Because the Doppler coefficient is higher at low temperatures the effective temperature T_{eff} for the Doppler effect is slightly lower than $\overline{T}$, but the difference is small. If the temperatures in the fuel element are in equilibrium, it can be shown that

$$\frac{T_{eff}}{\overline{T}} \simeq 1 - \frac{1}{3}\left(\frac{\overline{T} - T_o}{\overline{T}}\right)^2, \qquad 1.44$$

where T_o is the fuel surface temperature.

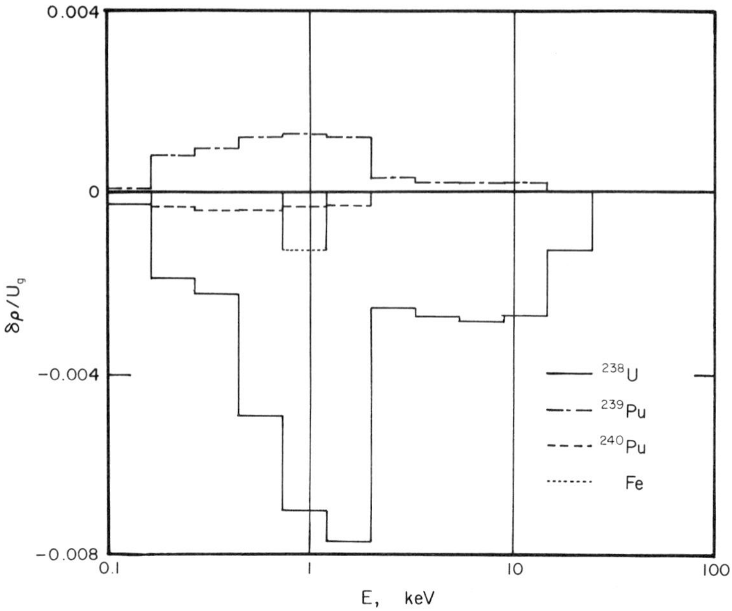

Fig. 1.19. Energy distribution of Doppler coefficient.

Temperature variation across the core can be taken into account by weighting changes with $\phi_g \phi^*_g$ as indicated in equation 1.29.

1.6.6 Power and Temperature Coefficients

In normal operation the temperatures of the various materials in the core vary in a regular way and it is useful to sum up their combined effects. There are various ways of doing this, one of which is to determine the "isothermal temperature coefficient". This is the rate of change of reactivity with temperature if the temperatures throughout the core vary by the same amount. It includes the effects of axial and radial expansion of the structure, of expansion of the sodium, and the Doppler effect. Because of the variation of the Doppler coefficient it depends slightly on the fuel temperatures. For a typical reactor it might be $- 2.5 \times 10^{-5}\ K^{-1}$ when the fuel is cold (i.e. the reactor is operating at very low power), and $- 1.5 \times 10^{-5}\ K^{-1}$ at normal operating temperature. It can be thought of as measuring the response of the reactor to a change in coolant inlet temperature when power and coolant flow-rate are held constant.

Another useful parameter is the power coefficient. This is the rate of change of reactivity with power, assuming the coolant flow-rate changes proportionately to the power and the inlet temperature is constant. The coolant and structure temperatures everywhere are thus constant so that the power coefficient depends only on the Doppler effect. A typical value of the power coefficient for a 2500 MW (heat) reactor is $-3 \times 10^{-6} (MW)^{-1}$.

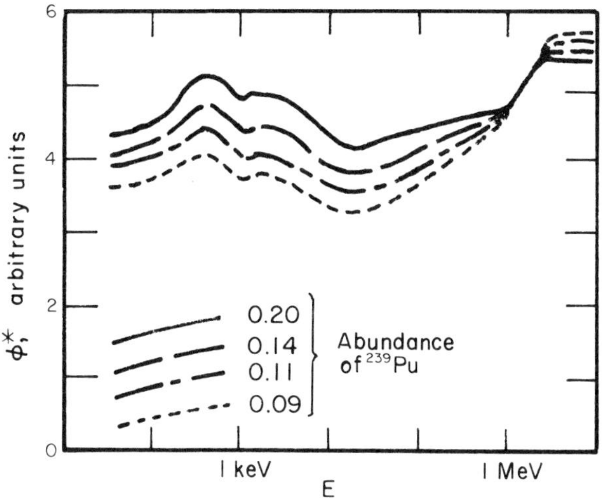

Fig. 1.20. Effect of enrichment on neutron importance.

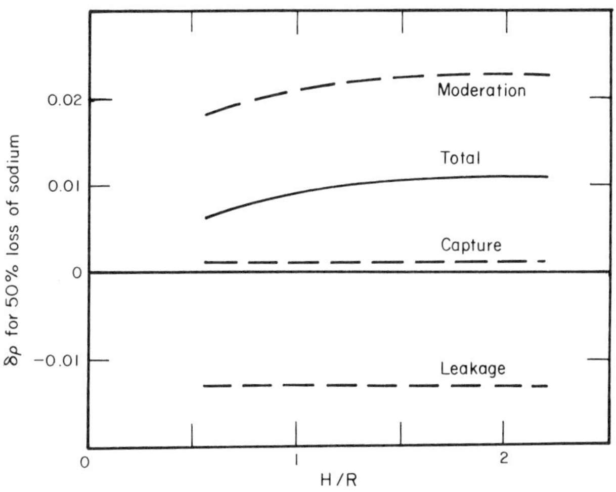

Fig. 1.21. Effect of core shape on sodium coefficient of reactivity.

1.6.7 Dependence of Doppler and Sodium Coefficients on Design

As we shall see in Chapter 5 the behaviour of the reactor in a serious accident may be strongly dependent on the Doppler effect and the reactivity change on loss of sodium. The consequences of certain accidents are likely to be less severe the more negative both of these are, and in fact they may form important limitations on the design. It is important therefore to know how design changes affect them.

An important factor is the ratio of fissile to fertile material in the core fuel. This is often called the "enrichment" of the fuel. Figure 1.20 shows the variation of ϕ^* with neutron energy for various values of the enrichment. As the enrichment increases (i.e. ^{239}Pu replaces ^{238}U) the importance of low-energy neutrons increases relative to that of high because of the reduced probability of ^{238}U fission and because high energy neutrons are more likely to leak from the smaller core because D is larger for fast neutrons than slow. As a result, the moderating component of the sodium coefficient becomes less positive as the enrichment is increased. At the same time the Doppler coefficient is made less negative by the reduced amount of ^{238}U.

The enrichment can be increased in a variety of ways. The height of the core can be reduced, or regions or layers of breeder can be incorporated within the core in various patterns (Okrent, 1964; Barthold, 1978). In all cases the sodium coefficient becomes more negative, the Doppler coefficient becomes less negative, the breeding gain increases, and the critical mass increases. The change in sodium coefficient is probably more important than the reduction in the Doppler coefficient from the point of view of safety, and cores of this type may be attractive as a result. The increase in critical mass outweighs the improved breeding, however, and together with the increased mechanical complexity gives an economic disadvantage which may well be more important. Figure 1.21 shows the effect of changing the height of a cylindrical core while keeping the volume constant.

Another way to reduce the sodium coefficient is to make the spectrum softer by incorporating more moderating material in the core. This must not, of course, be sodium but a solid moderator is suitable and beryllium oxide, BeO, has been proposed (Okrent, 1964). If some 10% of the core volume is occupied by BeO the sodium coefficient can be made negative and in addition the Doppler coefficient becomes more negative. This is in fact the only practical way both Doppler and sodium coefficients can be made more negative at the same time, but there are serious disadvantages in that the breeding gain is decreased and the critical mass increased, and there is a considerable economic loss in consequence.

Increasing the fraction of ^{241}Pu in the fuel also makes both Doppler and sodium coefficients more negative, but this is not really a useful design option because the effects are small and it is likely to be difficult to alter the composition of the plutonium in the fuel without a severe cost penalty. The effect on the sodium coefficient arises because the fission cross-section of ^{241}Pu rises more rapidly with decreasing energy than that of ^{239}Pu so the importance of low-energy neutrons is increased. The higher fission cross-section of ^{241}Pu in general means that the enrichment is lower so more ^{238}U is present and the Doppler coefficient is more negative.

Replacing oxide fuel by carbide makes the spectrum harder and so makes the sodium coefficient more positive. The Doppler coefficient becomes more negative because the enrichment is decreased.

1.7 SHIELDING

1.7.1 Calculation Methods

Diffusion theory, which works well for predicting the neutron flux in the reactor core and breeder, cannot be used for the neutron shield surrounding the breeder. It is incorrect for deep penetration of neutrons through an absorbing medium because the approximations used to obtain the relationship between neutron flux and current, which is the basis of the diffusion equation, are no longer valid. It does not predict correctly the attenuation of the small number of neutrons which penetrate without making any collisions at all. There are very few of these, of course, but deep in the shield they dominate the flux distribution.

The uncollided neutrons are predominantly very fast because most cross-sections are lower for fast neutrons than slow. The flux of uncollided or "removal" neutrons as they are known is attenuated exponentially, and varies as $\exp(-\Sigma_{rem}\,x)$ where x is distance and Σ_{rem} is the "removal" cross-section given by

$$\Sigma_{rem} = \Sigma_{total} - \bar{\mu}\Sigma_s, \qquad 1.45$$

where $\bar{\mu}$ is the mean cosine of the scattering angle. The $\bar{\mu}\Sigma_s$ term is an allowance for forward scattering, and is important at high energy where elastic scattering is often significantly anisotropic.

It is possible to calculate the flux of removal neutrons, together with the rate at which they make collisions, throughout the shield. The collision density is then used as a source term for a multigroup diffusion calculation, a small number of energy groups usually giving the required accuracy.

1.7.2 Shield Design

As fast neutrons penetrate deeply reactor shields contain where possible light moderating materials such as hydrogen or carbon to reduce the neutron energy together with absorbing materials such as boron or iron. The radial shield surrounding the breeder of the P.F.R., for example, is made of borated graphite blocks contained in thick steel tubes while the shield above the reactor contains hydrogenous resin and steel. Other designs omit the graphite from the radial shield. Concrete is a useful alternative hydrogenous material.

Figure 1.22 shows the attenuation of neutron flux through a typical reactor top shield. The high attenuation, by a factor of 10^{14} or so, is necessary to prevent activation of items of equipment such as coolant pumps or control-rod mechanisms to which access is needed for maintenance or repair. The radial shield has to protect the intermediate heat exchangers and prevent activation of the secondary coolant.

Because the attenuation is so high, however, irregularities in the shield can be very important. If for example there is a hole of area A through the shield the neutron current in the hole is attenuated as A/x^2, so that however small A is sooner or later the current through the hole becomes more important than the current through the bulk of the shield which attenuates as $\exp(-\Sigma_{rem}\,x)$.

In cases where there have to be holes through the shield to accommodate the control-rod drives or the pump shafts, for example, it is therefore important to make the clearances small and where possible to make sure they are not straight so that neutrons cannot penetrate without making any collisions. Even so such penetrations can dominate the performance of a shield.

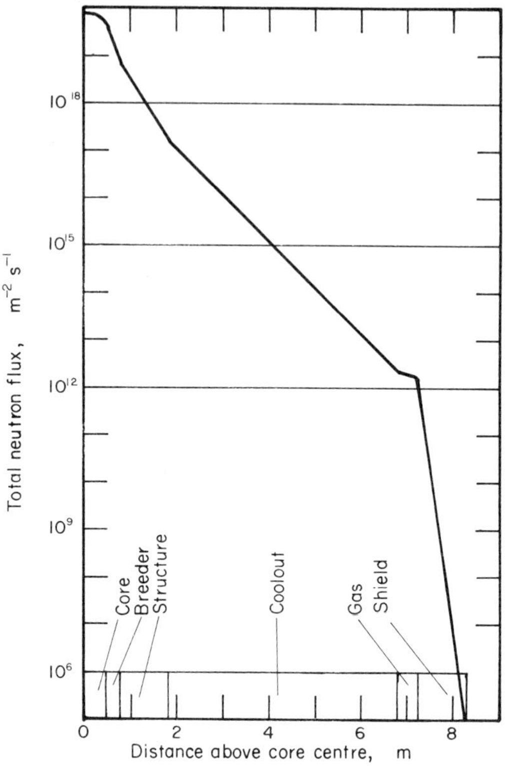

Fig. 1.22. Flux attenuation through sodium pool and shield.

CHAPTER 2

Fuel

2.1 INTRODUCTION

Most modern reactors - PWR, BWR, AGR, CANDU and SGHWR, as well as LMFBR - use oxide fuel, and much of this chapter applies as well to thermal reactors as to fast. The main difference is that new fast reactor fuel usually consists of a mixture of plutonium and uranium dioxides whereas thermal reactors use, at present, enriched uranium dioxide with a small amount of plutonium present after irradiation. Most water-cooled thermal reactors have fuel clad in zirconium alloy, but AGRs use stainless steel, and because of the similarity of the coolant temperatures there is much in common between the behaviour of AGR fuel and that of a sodium-cooled fast reactor.

There is an important difference, however. In all systems it is economically desirable to irradiate the fuel as long as possible before it is removed from the reactor. In a thermal reactor the limit is set by the loss of reactivity as fission products which absorb thermal neutrons accumulate. In a fast reactor the loss of reactivity is much smaller, especially if the internal breeding is good, and the limit is set by the need to be sure that the cladding will remain intact and not allow radioactive material to get into the coolant.

As a result the maximum burnup is very high in a fast reactor. The neutron irradiation accompanying this burnup damages the cladding so that the fuel has eventually to be removed and reprocessed before the cladding loses its integrity.

A very thorough account of reactor fuel in general, and in particular of oxide fuel for LMFBRs, is given by Olander (1976). It is particularly valuable for its treatment of the theory of many aspects of fuel behaviour.

2.2 FUEL TEMPERATURES

2.2.1 Temperature Distribution

In a cylindrically symmetrical fuel element, the temperature T_{fs} of the surface of the fuel is related to the temperature of the surrounding coolant T_c by

$$T_{fs} - T_c = \frac{q}{2\pi}\left(\frac{1}{R_s h} + \frac{\ln(R_s/R_f)}{K_s} + \frac{1}{R_f h_f}\right), \qquad 2.1$$

where q is the linear heat rating in the fuel element, h is the heat transfer coefficient between cladding and coolant, K_s is the thermal conductivity of the cladding, and h_f is the heat transfer coefficient between the fuel and the inside surface of the cladding. R_s is the outer radius of the cladding and R_f is its inner radius which we can also take to be the outer radius of the fuel because, as we shall see, at times the fuel is in close contact with the cladding, and even when it is not the gap between them is small.

The temperature distribution within the fuel itself is complicated by the fact that over the wide temperature range in question the variation of the fuel conductivity K_f has to be taken into account. If the fuel temperature at a distance r from the axis is $T_f(r)$, and if the fuel is solid and heat is generated uniformly throughout it, then

$$\int_{T_{fs}}^{T_f(r)} K_f(T)dT = \frac{q}{4\pi}\left(1 - \frac{r^2}{R_f^2}\right), \qquad 2.2$$

and the maximum fuel temperature T_{fm} is given by

$$\int_{T_{fs}}^{T_{fm}} K_f(T)dT = \frac{q}{4\pi}. \qquad 2.3$$

Under some circumstances, however, there is a cylindrical hole in the centre of the fuel. If this has a radius R_h and heat is generated uniformly throughout the annular fuel region the temperature at r is given by

$$\int_{T_{fs}}^{T_f(r)} K_f(T)dT = \frac{q}{4\pi(1-\alpha)}\left(1 - \frac{r^2}{R_f^2} + 2\alpha \ln\left(\frac{r}{R_f}\right)\right), \qquad 2.4$$

where $\alpha = R_h^2/R_f^2$. The maximum temperature is given by

$$\int_{T_{fs}}^{T_{fm}} K_f(T)dT = \frac{q}{4\pi}\left(1 + \frac{\alpha \ln \alpha}{(1-\alpha)}\right). \qquad 2.5$$

Comparison of equations 2.3 and 2.5, remembering that $\alpha < 1$, shows that for constant q the presence of a central hole reduces the maximum temperature.

2.2.2 Thermal Conductivity

Figure 2.1 shows the variation of $K_f(T)$ and $\int^T K_f(T')dT'$ with T for unirradiated UO_2 from 500°C to the melting point which is about 2800°C (Olander, 1976, p.127). At low temperature heat is conducted by the diffusion of phonons through the crystals. As the temperature increases the density of phonons increases and, as the behaviour of the crystal lattice is slightly non-linear, the probability of interaction between phonons increases, their mean-free-path decreases and as a result the conductivity decreases. At higher temperature still the conductivity increases again, possibly because heat transfer by radiation within the crystals becomes important, or because the specific heat capacity rises.

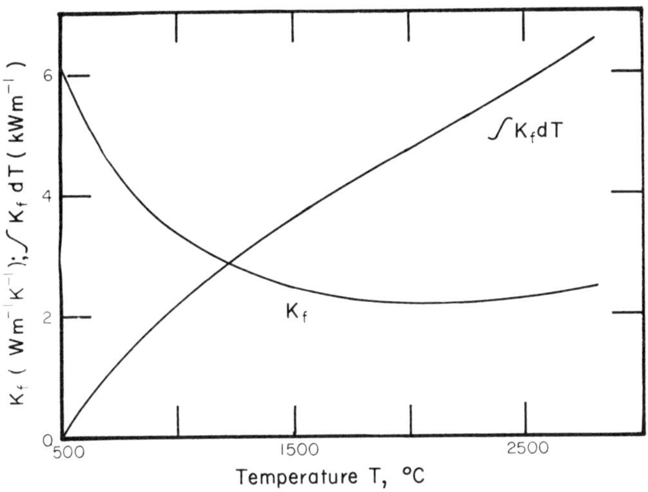

Fig. 2.1. Thermal conductivity of uranium dioxide.

The conductivity is affected by the chemical composition of the fuel, which can be described by a formula of the form $(Pu_aU_{1-a})O_{2+x}$, where a represents the plutonium fraction and x, which can be either positive or negative, the departure from the stoichiometric oxygen content. x is usually in the range $-.03 < x < .03$, although values as low as -.06 have been explored.

If there is no plutonium (a = 0) the effect of stoichiometry on conductivity is as follows. Below about 1400°C K_f decreases steadily as x increases, but above 1600°C K_f is a maximum for x = 0 and is reduced by departure from the stoichiometric composition in either direction. This seems to be because at high temperature an excess or a deficiency of oxygen, not exceeding a few per cent, is accommodated by vacancies or interstitials in the lattice. These defects reduce the phonon mean-free-path and hence the conductivity. At low temperatures, however, excess uranium ($x < 0$) is precipitated as the metal, and the resulting free electrons increase the conductivity.

The situation is quite different if plutonium is present. Uranium and plutonium have nearly the same ionic radius and stoichiometric UO_2 and PuO_2 form solid solutions in whatever ratio they are mixed (for all values of a). But if the mixed oxide is not stoichiometric with $x < 0$ the excess metal is accommodated by means of lattice defects for $x > -.02$ and by precipitation of Pu_2O_3 for $x < -.02$. Thus for mixed oxide with Pu fraction typical of a large fast reactor ($a \simeq 0.2$) the conductivity decreases as the composition departs from stoichiometric in either direction ($x < 0$ and $x > 0$) at all temperatures. For x = 0 the conductivity decreases slowly as a increases.

This picture is confused in practice by structural effects which can be much more important than the effects of composition. Principal among these is the effect of porosity, which reduces the effective conductivity. As explained below (section 2.3.2) the fuel may be manufactured with 10% or more porosity (i.e. 10% of the overall volume of the fuel may be occupied by voids). The effective conductivity is roughly proportional to 1 - p, where p is the porosity.

Estimation of the effect of porosity in an operating fuel element is made very difficult because, as explained in section 2.4.1 below, the pores move under the influence of the temperature distribution, so that the conductivity changes with time and in different ways in different parts of the fuel. Further confusion is provided by the cracks in the fuel which are formed under the influence of differential thermal expansion and which have a large but unpredictable effect on conductivity. Finally fission products, which themselves move through the fuel, and changes in crystal structure (section 2.4.1), affect conductivity in a manner which is not accurately known.

The maximum linear heat rating q is set by the requirement that the fuel should not melt. If the fuel surface temperature T_{fs} is 1000°C Fig. 2.1 gives

$$\int_{T_{fs}}^{T_{fm}} K_f(T)dT \simeq 4.3 \text{ kW m}^{-1}$$ for stoichiometric oxide. For oxide with excess metal (x = -.02 or so) the integral is rather greater partly because of the higher conductivity at low temperature and partly because the melting point is about 100 K higher.

Equation 2.3 shows that $\int K_f dT = 4.3$ kW m^{-1} implies $q \simeq 55$ kW m^{-1} if the fuel has no central hole. But if there is a hole with radius 20% of the fuel radius ($\alpha = .04$) equation 2.5 shows that q can be increased to about 63 kW m^{-1}. In fact the uncertainty of the effects of cracking and porosity and in the conductance between fuel and cladding (section 2.2.3) are such that q is usually limited to about 50 kW m^{-1}.

2.2.3 Conductance between Fuel and Cladding

A major uncertainty in predicting the fuel temperature is the value of the conductance across the interface or gap between the fuel and the cladding. This is h_f in equation 2.1.

If the fuel is manufactured in the form of pellets they are inserted into the cladding tubes with a radial clearance of some 50 μm or more when cold. This is reduced by the differential thermal expansion when the reactor is at power. The thermal expansion coefficient of the cladding (about 18 x 10^{-6}K^{-1}) is greater than that of the fuel (about 13 x 10^{-6}K^{-1}), but as the increase in the fuel temperature is greater the gap is reduced. The extent of the reduction is hard to predict, however, as the expansion of the fuel is irregular because it cracks (see section 2.4.2).

To increase h_f the fuel elements are usually filled with helium when they are manufactured because it has a high thermal conductivity of about 0.3 W m^{-1} K^{-1} at 600°C. If a helium-filled radial gap is 20 μm wide and symmetrical when at power, h_f is about 15 kW m^{-2} K^{-1}.

As irradiation proceeds the fuel swells and the gap becomes narrower until the fuel and the cladding are in contact, but because the surfaces are rough this does not mean that the effective width of the gap is zero. When two solid bodies touch each other contact is made only at a few places where extreme points on the two rough surfaces touch. The fraction of the apparent interface over which there is actual contact may be of the order of 1% and increases if there is a normal stress pushing the surfaces together. The effective thickness of the gas layer when the surfaces are in contact depends on their roughness and may be 0.5 μm if they are smooth, but cracks in the fuel increase it to an extent which is uncertain. Also the estimation of the area of solid to solid contact is uncertain.

A further large uncertainty is introduced by the changes in the composition of the gas in the gap. As irradiation proceeds inert fission-product gases, mainly xenon with some krypton, are released from the fuel (see section 2.3.4). Because of their high atomic weights they have much lower thermal conductivities than helium (.026 $W\ m^{-1}\ K^{-1}$ for krypton and .016 $W\ m^{-1}\ K^{-1}$ for xenon at 600°C). The rate at which they are released from the fuel is not known accurately, and as they reduce the conductivity of the helium substantially by diluting it they introduce another uncertainty in the conductance.

In principle it is possible to estimate h_f experimentally, either by examining fuel after irradiation and deducing the temperature distribution from the observed changes in the structure, or by measuring the fuel temperature during operation by inserting thermocouples into it. Either method involves large uncertainties however, mainly because the conductivity of the fuel itself is not known accurately as explained above.

As a result it is usually best to assume an approximate value of h_f of about 5 or 8 $kW\ m^{-2}\ K^{-1}$, and to recognise that it is very uncertain and that the actual value may be up to a factor of two different (Olander, 1976, p.139). For $R_f = 3$ mm and $q = 50\ kW\ m^{-1}$, $h_f = 6\ kW\ m^{-2}\ K^{-1}$ gives an interfacial temperature difference between fuel and cladding of 440 K, but the actual value may be anywhere between 200 K and 800 K.

2.3 FUEL DESIGN AND MANUFACTURE

2.3.1 Fuel Swelling

Almost every fission event turns one atom (of uranium or plutonium) into two (of fission products). Some of these, as we shall see (section 2.3.4), are gaseous or volatile and mostly lost from the fuel, but the majority are solid. They are retained within the fuel either as interstitial atoms in the crystals, as inclusions of a distinct solid phase, or, in the case of gaseous fission products, as small bubbles within the crystals or on the grain boundaries.

Very roughly all atoms occupy about the same volume (of the order of $10^{-29}\ m^3$) in solid and liquid phases so the multiplication of the number of atoms means that the volume increases. Loss of some fission products from the fuel tends to mitigate the increase, but on the other hand retention of bubbles of gaseous fission products tends to enhance it. The nett result is a fractional increase in the volume of the solid fuel of about 0.8B, where B is the burnup.

This swelling is almost inexorable and if the fuel fits closely inside the cladding when it is new the cladding is forced to strain to accommodate it. Thus for example 10% burnup would imply about 3% normal strain in the cladding in both the circumferential and (because the friction between fuel and cladding may not allow relative motion) axial directions. As pointed out in section 3.3.4 the ductility of the irradiated cladding may be too low to accommodate it.

The primary purpose of the cladding is to retain the radioactive fission products, and if more than a very few fuel elements crack because of the swelling of the fuel inside them the design is unacceptable. The design has to be optimised to maximise the burnup without causing failure of the cladding. The fuel has to be removed from the reactor when it has reached a burnup beyond which trial irradiation has shown that an unacceptable number of failures will occur.

It is usual to allow for the effects of fuel swelling by leaving space for the fuel to swell without straining the cladding. In some cases this has been done by manufacturing the fuel in the form of pellets with a central hole so that the increase in volume can be accommodated by filling or partially filling the hole. Alternatively and more usually the fuel is manufactured as porous pellets, the pores occupying some 10-20% of the overall volume, and the swelling is accommodated by filling the pores.

It is by no means certain, however, that either pores or a central void work in the manner intended to accommodate swelling. As explained in section 2.4.1 pores migrate during operation and a central void is often formed even when one is not present to start with. In addition the fuel creeps (see section 2.4.3) so that there is a tendency for stress between fuel and cladding to be relieved by strain of the fuel as well as of the cladding, and irradiation-induced reduction of the density of the cladding (section 3.3.2) also helps to accommodate fuel swelling. In fact the picture is so confused that, although the strain of the cladding of an irradiated fuel element can be measured, it is not easy to predict.

2.3.2 Manufacturing Processes

In most reactors the fuel is in the form of sintered pellets which are made from a mixture of UO_2 and PuO_2 powders in the correct proportion. This can be obtained either by making pure UO_2 and PuO_2 powders and mixing them or alternatively by mixing nitrate solutions in the correct proportions and forming the mixed oxide by simultaneous precipitation and subsequent reduction.

The advantage of the latter technique, which is known as "co-precipitation", is that the resulting mixture is more homogeneous. If pure oxide powders are mixed there is a risk of hot spots due to high local plutonium concentration. There is also the danger that if the power increases very rapidly in an accident the effect of the negative Doppler coefficient of reactivity from the ^{238}U (see section 1.6.5) may be reduced if the delay in transferring heat from the plutonium-rich regions, where most of the fissions occur, is significant.

The oxide power is then mixed with a binder or plasticiser, usually an organic compound, which allows the powder to be pressed into the required shape. The pellets are then sintered to turn them into the required ceramic form. During the sintering process the binder is broken down and driven off. The porosity of the finished pellets can be controlled by careful choice of the nature and quantity of the binder and the sintering time and temperature, and can be made as low as 2% if required.

Finally the pellets may have to be ground to the correct size. This is an expensive process because it produces plutonium-bearing dust which is hazardous and therefore has to be controlled carefully, and it is probable that eventually precise control of the sintering process will produced pellets of the required dimensions within a sufficiently close tolerance.

An alternative manufacturing technique is the "vibro-compacted" or "vipac" method. In this small sintered particles of oxide are placed in the cladding tube which is then vibrated until they are compacted to the required density. To obtain a high packing fraction particles of at least two sizes have to be used, and typical diameters are 800 and 80 μm.

A disadvantage of "vipac" fuel is that even if the porosity of the individual particles is as low as 3% the overall porosity of a 6 mm diameter fuel element made with 800 and 80 μm particles cannot be reduced below about 18%. To achieve higher overall fuel density a third, even smaller, particle size would be required, which would increase the cost of manufacture significantly. Another disadvantage is that if the small particles, or "fines" as they are called, contain plutonium they are hazardous in manufacture because of the risk of inhalation. If the plutonium is confined to the larger particles the hazard is reduced but the disadvantages of possible hot-spots and a delay in the Doppler effect are incurred. In spite of these disadvantages the "vipac" method is attractive because of its simplicity and possible cheapness, and may be more widely used in future.

Whether pellet or "vipac" fuel is used it has to go through the sintering process, and at this stage the oxygen content can be controlled. If the sintering is carried out in a reducing atmosphere, containing hydrogen for example, the oxygen content can be reduced to a sub-stoichiometric level (i.e. $x < 0$).

During sintering the majority of the binder is removed but unless care is taken traces of hydrogen and carbon may remain which can have deleterious effects on the cladding.

2.3.3 Stoichiometry and Oxygen Potential

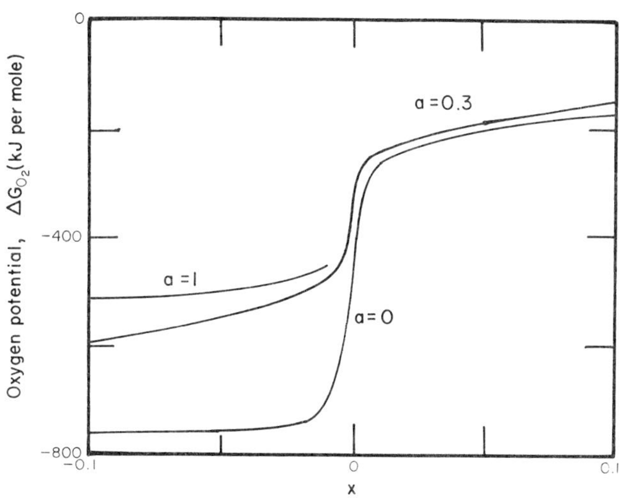

Fig. 2.2. Oxygen potential of $(U_{1-a} Pu_a)O_{2+x}$ at 1500°C.

Corrosion of the cladding by the fuel is discussed in section 2.4.7. It is strongly affected by the oxygen potential in the fuel and therefore by the stoichiometry. Figure 2.2 shows the variation of the oxygen potential of $(U_{1-a} Pu_a)O_{2+x}$ with x at constant temperature for various values of a (Olander, 1976, p.148). As is to be expected it rises with x, but also it is higher for

PuO_2 (a = 1) than for UO_2 (a = 0) because uranium adopts higher valancy states than plutonium. The steep rise in the oxygen potential of UO_{2+x} as x increases through zero is important. The variation with temperature is shown in Fig. 2.3. Oxygen tends to migrate to the cooler parts of the fuel if $x < 0$, but if $x > 0$ the tendency is considerably reduced.

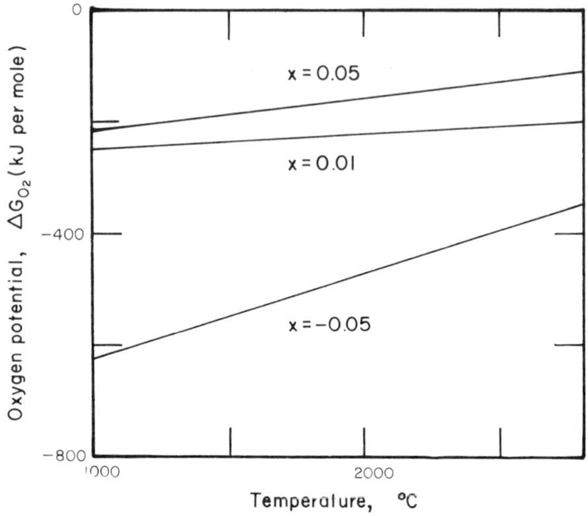

Fig. 2.3. Effect of temperature on the oxygen potential of $(U_{0.7}\,Pu_{0.3})O_{2+x}$.

The effect of burnup is very complicated because of the wide range of elements formed as fission products (Findlay, 1974). Each fission releases two oxygen atoms, some of which go to oxidise the fission products, such as zirconium, strontium, barium, and the rare earths, which have a strong affinity for oxygen. The number of oxygen atoms required for this depends on the yields of the various fission products, and the yields differ for fission of different nuclides.

Fission of uranium yields more zirconium and strontium (which form oxides) and less ruthenium and palladium (which do not) than fission of plutonium. It so happens that as a result the average requirement to oxidise the fission products is for slightly more than two oxygen atoms per uranium fission and for slightly less than two per plutonium fission.

The effect is shown in Fig. 2.4. In typical reactor core fuel far more fissions occur in ^{239}Pu than in ^{238}U, so oxygen is steadily released and the oxygen potential of the mixture of fuel and fission products rises. Just as in Fig. 2.2 there is a particularly sharp rise at the point in burnup where the oxygen content becomes stoichiometric for the mixture.

The oxygen potential does not rise very high, as Fig. 2.4 indicates. An effective upper limit is set by molybdenum. The Mo/MoO_2 system has an oxygen potential very similar to that of the fuel, and as the yield of molybdenum is high (about 0.06

atoms for each fast fission of either ^{239}Pu or ^{238}U) it forms a buffer. When the stoichiometric ratio is reached, oxygen released by further fissions is taken up by oxidising molybdenum (Olander, 1976, p.181).

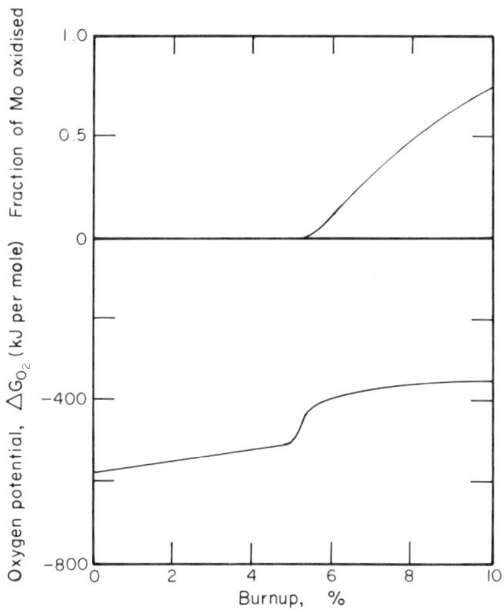

Fig. 2.4. Effect of burnup on the oxygen potential and oxidation state of molybdenum for $(U_{0.85}\,Pu_{0.15})O_{1.96}$.

2.3.4 Fission-Product Gas Release

TABLE 2.1 Fission Yields of Long-Lived Isotopes of Inert Gases (Tamplin, 1963)

Isotope	Half-Life	Fission Yield (atoms per fission)
Krypton 83	Stable	.005
" 84	"	.010
" 85	10.3 y	.003
" 86	Stable	.020
Xenon 131	Stable	.029
" 132	"	.043
" 134	"	.080
" 136	"	.064

Various isotopes of krypton and xenon are produced by fission. The cumulative yields of the stable and long-lived isotopes are shown in Table 2.1. Most of them are formed not immediately but as the end-products of beta-decay chains (Tamplin, 1963). The half-lives of most of the decay processes are of the order of a few minutes, and in only two cases (^{131}Xe and ^{132}Xe) of a few days. The total yield is about 0.04 atoms of krypton and 0.22 atoms of xenon per fission.

This is a very large quantity. If fuel of density 10^4 kg m^{-3} undergoes 10% burnup and the krypton and xenon generated exist as gas at 400°C and atmospheric pressure they occupy 53 times the volume of the fuel. Alternatively if they are confined in a volume equal to that of the fuel, they exert a pressure of 5.3 MPa at 400°C or 6.8 MPa at 600°C.

The krypton and xenon are not formed as gases, however, but as single atoms in solution in the fuel. The solubility is very low so the solution is soon super-saturated and the atoms are precipitated to form bubbles. They enter the bubbles by thermal diffusion (and therefore at a rate which depends on temperature), but they may also leave the bubbles and return to the fuel by the process known as fission-induced resolution.

This comes about as follows. If a fission occurs close to a gas bubble one of the fission-fragments may enter it. The fission-fragment may have 100 MeV or more of kinetic energy, and if it collides with a gas atom it can easily impart enough energy to it for it to re-enter the surrounding fuel crystals. In effect the gas atom is knocked out of the bubble and forced back into solution.

The size and rate of growth of a bubble depend on the rates at which it gains and loses gas atoms. Bubbles within a grain gain by diffusion and also by sweeping up atoms or other bubbles as they move through the grain (see section 2.4.1 for the mechanism of this process) and lose by resolution. They are typically spherical, with a diameter of the order of 50 nm. Bubbles at the grain boundaries, however, grow as the grain boundaries move. The hotter part of the fuel is continually being recrystallised, and as one grain grows at the expense of a neighbour the grain boundary sweeps through the fuel collecting gas atoms and bubbles as it goes. This is a very effective growth mechanism and as a result the grain boundary bubbles are much larger. They are sometimes flat or lenticular in shape and up to 20 μm across. Figure 2.5 shows typical gas bubbles.

As the grain boundary bubbles grow they link up and eventually form channels by which the gas escapes from the fuel. The fraction of the gas which is released in this way depends in a complex manner on burnup and temperature. Because the process is so complicated it is usual to assume when designing the fuel elements that all the krypton and xenon generated will be released as gas.

In reality, however, some 10 or 20% is retained in the fuel, and this may become very important in the course of an accident. If the fuel becomes overheated gas retained in the colder parts may suddenly be released and affect the subsequent course of events (see section 5.4.3). Cox and Homan (1970) make an approximate estimate of the amount available by assuming that 2% of the gas generated in fuel at temperatures above 1800°C is retained, 50% in fuel between 1800°C and 1400°C, and 70% in fuel below 1400°C.

2.3.5 Sealed or Vented Fuel

Given that allowance has to be made for this large quantity of gas to be given off by the fuel during irradiation the next question is whether it should be retained within the fuel element or released from it. If it is retained space has to be provided for it. If a volume equal to that of the core fuel is provided, after 10% burnup the pressure inside the fuel element may be some 5 - 7 MPa as we have seen. If a smaller volume is provided the thickness of the cladding has to be increased, which reduces the breeding gain. If a larger volume is provided the

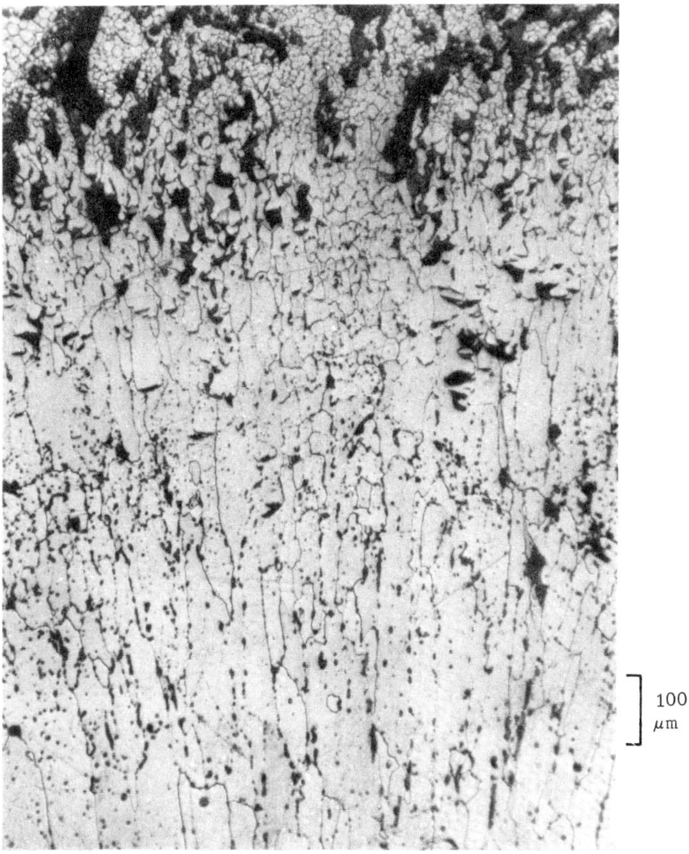

Fig. 2.5. Gas bubbles in irradiated fuel.
(0.8% burnup at 50 kW m^{-1})

fuel element has to be longer, the reactor vessel and the fuel handling equipment have to be larger and therefore more expensive, and a greater pressure drop is needed to circulate the coolant.

If the gas were released from the fuel elements there would be considerable advantages in the resulting reduction of the height of the reactor vessel and possibly of the cladding thickness. There are, however, important disadvantages, the worst of which is the difficulty of keeping the coolant out of the fuel elements.

If sodium comes into contact with the fuel, under some circumstances a chemical reaction takes place and a mixture of sodium uranate and plutonate (Na_3UO_4 and Na_3PuO_4) is formed. These compounds have a low density so the fuel swells when the reaction takes place and may burst the cladding (Olander, 1976, p.555). If the gas is to be released, therefore, each fuel element has to be fitted with a vent which will allow the gas to leave but will prevent coolant entering even when the reactor is shut down and cooled below its operating temperature so that the pressure inside the fuel elements is less than that outside. The vent has to be cheap enough to be made in large numbers and yet very reliable. Several designs have been tried but none has yet been put into widespread use.

A second disadvantage of vented fuel is that it inevitably allows the release of radioactive materials into the coolant. Apart from the isotopes listed in Table 2.1 various short-lived isotopes of krypton and xenon such as ^{90}Kr (half-life 33 s), metastable ^{83}Kr (114 min) and ^{133}Xe (5.3 d) appear as fission products. To prevent or minimise release of these isotopes a device has to be incorporated which delays the release of the gas until they have decayed, because the presence in the coolant of their decay products such as ^{90}Sr of which ^{90}Kr is a precursor is very undesirable.

Of course nothing can be done to prevent the release of ^{85}Kr from vented fuel and it is very hard to guarantee that other volatile fission products such as ^{131}I (8 d) and ^{137}Cs (29 yr), of which the yields are high, will not be released with the inert gases.

2.3.6 Fuel Element Design

An important choice facing the designer of sealed fuel elements is whether to place the gas storage volume, or plenum as it is usually called, above or below the core. If it is below, surrounded by inlet coolant at about 400°C, it can be smaller than if it is above at 560°C or so. But if for some reason the plenum should burst or leak the gas might be released rapidly and pass upwards, displacing coolant from the core and possibly causing overheating and damage. Currently opinion is divided as to which is the better arrangement.

Between the core fuel and the plenum there is ideally an axial breeder region containing natural or depleted UO_2. Provision has to be made to allow the gas from the core to pass through the breeder to the plenum. The axial breeder on the opposite side of the core (i.e. above the core if the plenum is below) can be incorporated in the same fuel elements with the core, or alternatively it can be made in the form of separate fuel elements which, because of the low power density in the breeder, can have a larger diameter than the core elements. The advantage of the latter arrangement is that the resistance to coolant flow is reduced. It has not been used often, however, because of the increased complexity.

Figure 2.6 shows a typical core fuel element which incorporates both axial breeders, has the fuel in the form of pellets, and has the plenum below the core. The pellets are retained in position by a spring as shown.

The diameter of the fuel element is determined by heat transfer and manufacturing cost considerations (section 3.2.1). The thickness of the cladding has to be sufficient after allowing for corrosion both on the outside (section 3.3.5) and the inside (section 2.4.7) to withstand the stresses due to fuel swelling and fission-product gas. If 316 stainless steel is used the thickness is usually about 0.3 or 0.4 mm.

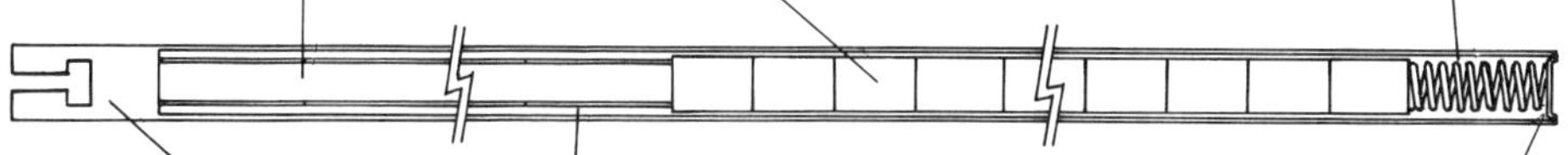

Fig. 2.6. Typical fuel element.

The radial breeder fuel elements are usually similar in form to those of the core but with larger diameter. Even at the end of several years irradiation the power density in the breeder adjacent to the core is only 20% or so of the power density at the core centre. The breeder elements can therefore have more than twice the diameter of the core elements before the limiting linear heat rating is reached.

2.3.7 Reprocessing and Design

It is not within the present scope to consider the reprocessing of the irradiated fuel in detail but there are some ways in which the design has to take into account the needs of the reprocessing plant, and to understand these some consideration has to be given to the processes involved. The main points are discussed by Chamberlain and others (1977).

After irradiation the fuel is stored for six months or so until the radioactivity has decayed to a manageable level. At the end of this time a typical fuel sub-assembly may have a radioactive source strength of some 1.5×10^6 Curies, and be liberating some 10 kW of heat. It is then taken to a reprocessing plant.

In most cases the chemical separation of fission products, plutonium and uranium is carried out in solution and so the fuel has to be dissolved in nitric acid. Before this is done, however, the wrappers and the end-sections of the fuel elements have to be removed. This is necessary because any of the steel which goes with the fuel into the nitric acid becomes contaminated with plutonium and therefore has to be protected very carefully when it is disposed of. The smaller the amount of steel contaminated in this way the less expensive the disposal.

It is therefore necessary to design the fuel elements and subassemblies, and especially the fuel-element supports (grids or wire wraps), so that the elements can be removed easily from the wrappers. The disassembly process must be very simple because it has to be done remotely, and maintenance of the machinery involved or correction of any faults that arise during the disassembly may be extremely difficult and time-consuming.

It is desirable to reprocess the fuel with the minimum delay while the decay heating is still significant. This means that the fuel has to be designed so that it can be cooled during disassembly, and the cooling has to continue reliably even if the disassembly machinery breaks down.

As well as the wrappers, the end fittings and the fission product gas plena are removed from the fuel elements so that only the section of cladding containing the core and axial breeder fuel is left. The fuel elements are then dissolved in small batches in such a way that criticality is not possible.

$(U,Pu)O_2$ containing more than about 40% plutonium does not dissolve readily in nitric acid. It is therefore necessary to ensure that the fuel is mixed uniformly during manufacture for particles of fuel with a high plutonium content might not dissolve and would require special treatment and possibly cause difficulty at some later stage in the process. It is not difficult to ensure uniformity of co-precipitated fuel, but if pellets are formed from mixed UO_2 and PuO_2 powders care has to be taken that the grains are very small and are mixed thoroughly.

In "vipac" fuel, if all the plutonium is in the larger granules (section 2.3.2) the plutonium concentration in the large granules has to be greater than the average in the fuel as a whole, and if after irradiation (when the plutonium concentration is still higher) it exceeds 40% difficulty will be experienced in dissolution.

2.3.8 Carbide Fuel

Although this chapter is mainly a description of oxide fuel it is interesting to consider briefly the main alternative, which is carbide, to show how the disadvantages of oxide may be overcome in the future.

The main advantages of fuel consisting of a mixture of UC and PuC are that its density is high, its thermal conductivity is high, and it is chemically compatible with sodium. The high density improves the breeding gain as compared with that of an oxide-fuelled reactor.

The high thermal conductivity means that the fuel elements can be larger, and therefore cheaper. Compatibility with sodium has the great advantage that the gap between fuel and cladding can be filled with sodium so that the conductance is very high. The result is that a fuel element can have a larger diameter or a higher linear heat rating, and still have a low central temperature.

Many advantages stem from the low fuel temperature. The fission product gas is released in much smaller quantities so that a smaller gas plenum is needed and the cladding stress due to the gas pressure is small. (There may be a parallel disadvantage in that a lot of gas is available to be released if the fuel should be accidentally overheated.) Carbide fuel may not swell under irradiation to the same extent as oxide so that another source of cladding strain is removed.

There are two possible disadvantages. If the gas which is released should accumulate as a bubble in the sodium bond between fuel and cladding there could be considerable local overheating, which might damage the fuel. There is also a danger of carburisation of the cladding if the sodium transports carbon to it from the fuel. It seems, however, that this can be prevented if the carbon potential of the fuel is kept low. These disadvantages can be avoided if the gap is filled with gas as in oxide fuel, but this increases the fuel temperature so that the other advantages are reduced.

2.4 BEHAVIOUR DURING IRRADIATION

2.4.1 Recrystallisation

During irradiation the crystal structure of the fuel is changed almost completely. Figure 2.7 shows a polished and etched cross-section of typical pellet fuel after irradiation and Fig. 2.8 shows a similar axial cross-section of "vipac" fuel. Throughout most of the cross-section the original particulate structure of the "vipac" fuel has completely disappeared.

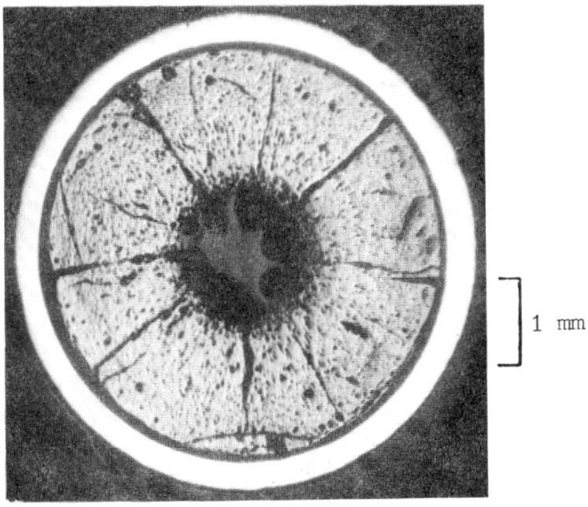

Fig. 2.7. Cross-section of fuel element irradiated to 7.2% burnup at 40 kW m^{-1}. (The grey object in the central void is resin used to fix the fuel in place when it was sectioned.)

The most striking change is that a hole appears in the centre. This is surrounded by a region of high density in which the grains are long and narrow and lie along radii of the cylinder. This is known as the "columnar grain" region. Next is a region where the grains are larger than in the original material but are oriented at random. This is called the "equiaxial grain" region. Finally, in the outermost part of the fuel it retains its original structure, and this is called the "unrestructured" region. These three distinct regions are shown in Fig. 2.9.

Observations made after various periods of irradiation show that this basic structure is set up very quickly - within an hour of achieving the full power density. Thereafter the pattern of equiaxial and columnar grains spreads outwards from the centre but at a rate which decreases very rapidly. The higher the linear rating the greater the radii of the restructured regions, and for most of the life of the fuel it is a fair approximation to regard the outer edge of the columnar grain region as being close to the 1800°C isotherm, and that of the equiaxial region to the 1600°C isotherm.

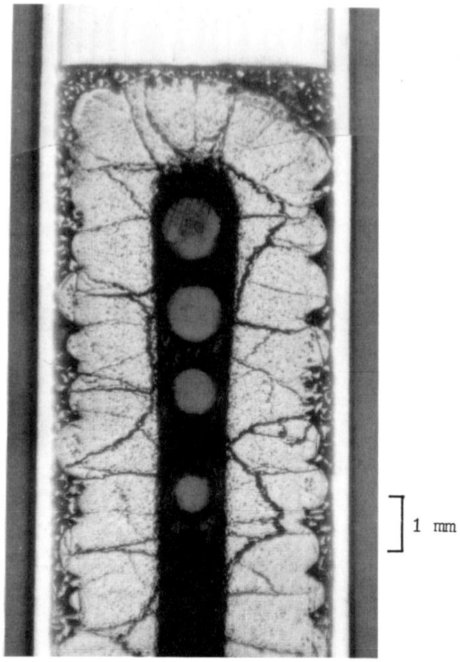

Fig. 2.8. Axial cross-section of a "vipac" fuel element irradiated to 4.6% burnup at 31 kW m^{-1}. (The round objects in the central void are bubbles in the resin used to fix the fuel in place when it was sectioned.)

As far as is known the restructuring takes place as follows (Olander, 1976, p.275). At intermediate temperatures the sintering which started during manufacture continues. The pores between the grains tend to coalesce and the individual grains grow to form the equiaxial grain region.

At higher temperatures the pores become mobile and move up the temperature gradient. There are two mechanisms for this: surface diffusion whereby atoms of fuel move round the pore from the hot to the cold side, and volume diffusion whereby atoms evaporate from the hot side and condense on the cold. The result is both that the pore moves towards the region of higher temperature and also that the fuel is recrystallised as the pore moves through it. The atoms deposited on the cold side of the pore form a single new crystal, relatively free from imperfections, occupying the volume swept out by the pore as it moves.

The pores move to the centre - the hottest part - of the fuel and there form the central void, while the density of the columnar grain region increases to something close to the theoretical value, which is that of a single crystal. The speed with

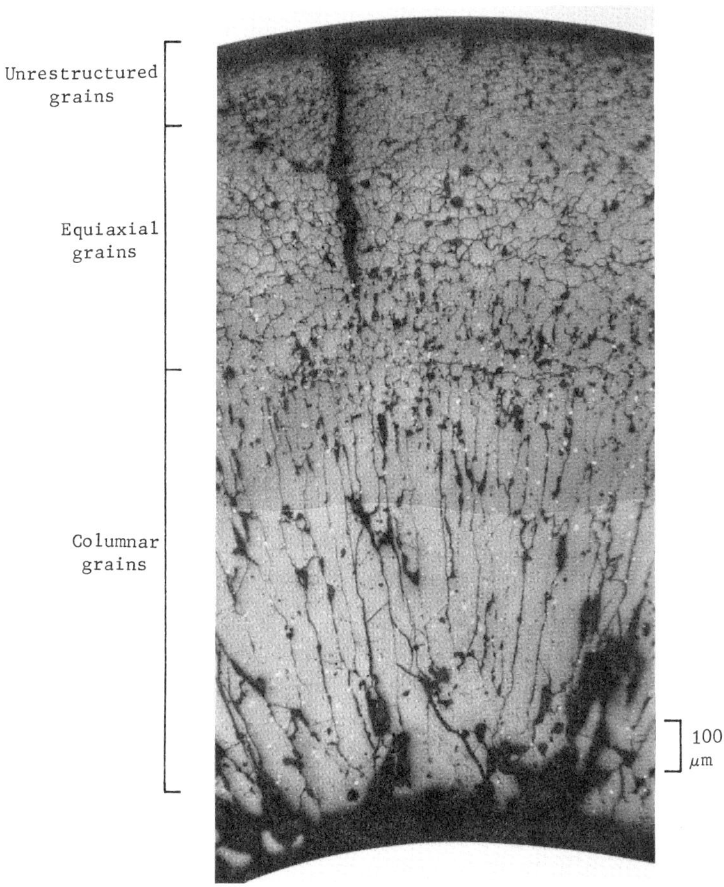

Fig. 2.9. Cross-section of fuel irradiated to 8% burnup at 40 kW m^{-1}, showing recrystallisation.

which the pores move depends strongly on temperature so as in the case of the equiaxial region the outer boundary of the columnar region moves quickly to start with and then very slowly.

Initially it is the pores incorporated in the material when it is manufactured which move in this way but later it is the bubbles formed by the accumulation of fission-product gas on the grain boundaries (section 2.3.4). As a result the columnar grain region is continually being recrystallised.

2.4.2 Cracking

The temperature difference of 1500 K or more between centre and surface of the fuel induces substantial thermal stresses which cause cracks. Fig. 2.7 shows a cracking pattern typical of those seen in irradiated fuel. It is important to realise, however, that the pattern of cracks seen when the fuel is cool is quite different from that present during irradiation. The conjectured development of cracks in pellet fuel is shown in Fig. 2.10.

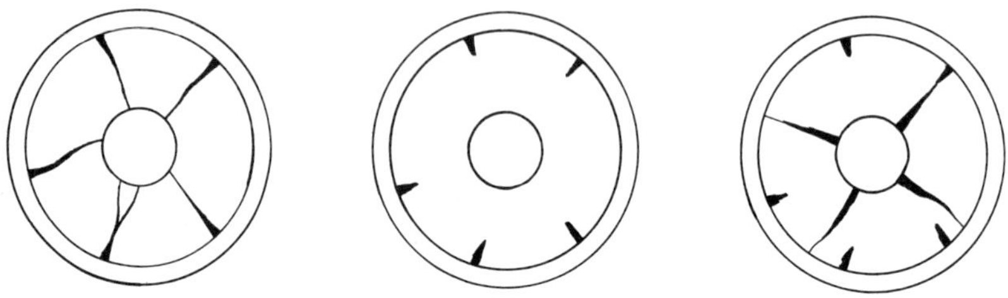

Fig. 2.10. Development of cracks in fuel during irradiation.

When the fuel first generates power and experiences a temperature difference stresses are set up as indicated in Fig. 2.10a. The tensile hoop stress at the outside exceeds the rupture stress and radial cracks are formed. In the centre, however, cracks are very quickly healed by the recrystallisation process.

After a certain amount of irradiation, usually about 0.1% burnup, the fuel has swollen to touch the cladding. Subsequent swelling is resisted by the cladding and the fuel is subject to a compressive stress. Under the influence of this it creeps (section 2.4.3), rapidly near the centre, very slowly at the periphery. As a result the cracks close at the centre but not at the outside. When they have closed sintering or recrystallisation heals them. Thus after prolonged irradiation the centre of the fuel is free of cracks while the periphery retains cracks which tend to taper inwards (Fig. 2.10b).

When the reactor is shut down the fuel cools, the centre contracts more than the periphery, and new radial cracks are formed which are wide at the centre and taper towards the outside (Fig. 2.10c). These cracks are prominent in pictures of fuel cross-sections, such as Fig. 2.7, but it should be remembered that they were not present when the reactor was operating.

2.4.3 Thermal and Irradiation Creep

Thermal creep in the fuel is usually described by an empirical relationship of the form

$$\dot{\epsilon}_{th} = A\sigma^n e^{-B/T}, \tag{3.6}$$

where σ is the stress, $\dot{\epsilon}_{th}$ is the strain rate, T is temperature and A, B and n are constants. Usually $n = 1$. B is about 4×10^4 K, and typical results for stoichiometric UO_2 $(x = 0)$ give $A \simeq 2 \times 10^{-4}\ s^{-1}\ Pa^{-1}$. If the fuel is not stoichiometric it is harder, while mixed oxide is usually found to be softer.

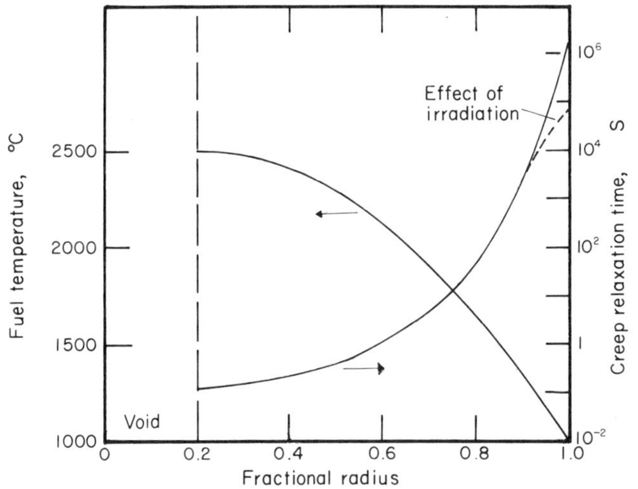

Fig. 2.11. Creep relaxation time in a fuel element generating 45 kW m^{-1}.

The resulting relaxation time for thermal stress is indicated in Fig. 2.11. It is very short in the centre of the fuel, and over times of the order of seconds or more this part of the fuel can support only hydrostatic pressure. The outer part of the fuel on the other hand is quite rigid.

A second source of creep strain is due to the irradiation itself. The effect of a fission event is to melt a small volume of the fuel. Any shear stress in the melted region is therefore transferred to the surrounding material which strains a little more as a result. Later another fission event will melt another region causing additional strain, and so on. The result is a steady strain-rate proportional to stress and fission rate density but only weakly dependent on temperature. Various experimental correlations have been proposed (Olander, 1976, p.353; Perrin, 1972), typically of the form

$$\dot{\epsilon}_f = A' f \sigma \tag{3.7}$$

with $A' \simeq 1.8 \times 10^{-36}$ m^3 Pa^{-1}. f is the fission rate density in m^{-3} s^{-1}, so that with σ in Pa equation 3.7 gives $\dot{\epsilon}_f$ in s^{-1}.

The effect of irradiation creep is also shown in Fig. 2.11. The relaxation time in the outer part of the fuel is substantially reduced by it but is still hard on a timescale of days or weeks.

It is of course impossible to observe what happens to the fuel during irradiation, but it is probably more-or-less as follows. After the fuel has swollen to touch the cladding, the cladding exerts a compressive normal stress on the surface of the fuel. The resulting stress distribution is shown in Fig. 2.12. The fuel is incapable of exerting hoop stress in the central region which is soft and in contact with the central void and therefore stress-free. It is also incapable of exerting hoop stress at the outer edge where there are radial cracks. Thus the compressive surface stress is borne by a compressive hoop stress in a narrow ring of fuel just at the root of the cracks.

In steady operation the load-bearing ring moves slowly outwards as more of the fuel has time to relax the imposed stress and allow the cracks to close. But if there are changes in reactor power and therefore in the fuel temperature distribution the development is interrupted. If the power rises, for example, the cracks re-open to some extent and the load-bearing ring is moved inwards.

2.4.4 Interaction between Fuel and Cladding

The fuel swells at a roughly constant rate throughout irradiation and has usually closed the gap and come into contact with the cladding after about 0.1% burnup. The cladding also swells (section 3.3.2) but only slowly at first, more rapidly later, and at a rate which depends strongly on temperature. The result is that for some cladding materials and in some parts of the core the swelling rate of the cladding may eventually exceed that of the fuel and tend to reduce the stress between the two.

While cladding and fuel are in contact and there is a compressive stress between them as shown in Fig. 2.12 both creep. Swelling of the central part of the fuel inside the load-bearing ring is accommodated by expansion into the central void. Swelling of the outer part is accommodated partly by the swelling of the cladding.

The cladding strain is not always uniform and in some cases may be concentrated by cracks in the fuel, and this, rather than uniform strain, may determine the maximum burnup to which the fuel can be subjected.

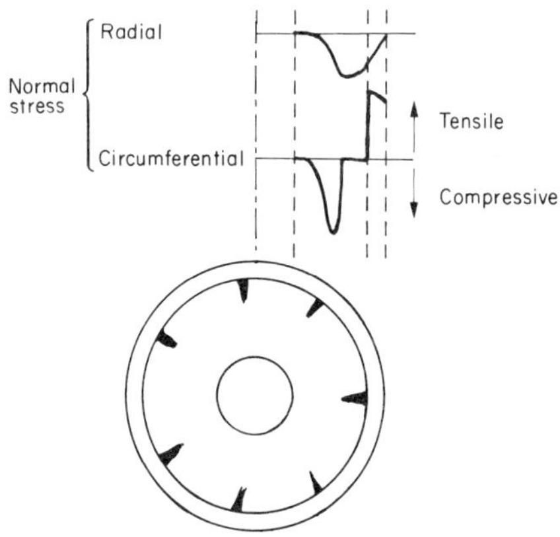

Fig. 2.12. Stress distribution in a fuel element during irradiation, after the fuel has made contact with the cladding.

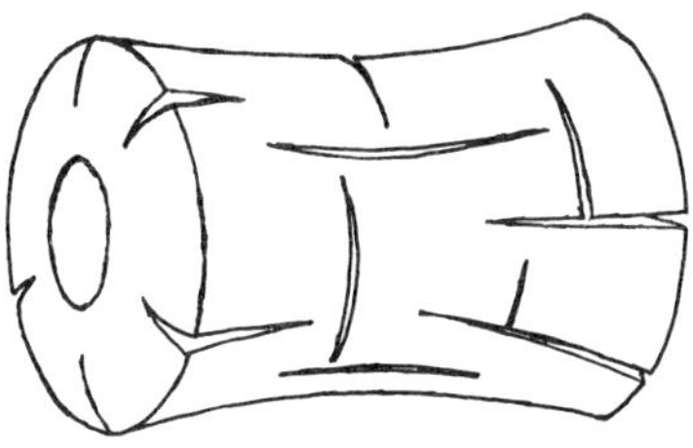

Fig. 2.13. Distortion of a fuel pellet during irradiation (exaggerated).

Sometimes the pellets themselves can cause non-uniform strain of the cladding. During irradiation a single pellet tends to distort into the shape shown, very much exaggerated, in Fig. 2.13. The ends are displaced outwards as shown and can cause ridges round the cladding giving it the appearance of a bamboo cane. These ridges are not usually permanent as the pellets soon fuse together, but they are sometimes observed in fuel which has been subject to abnormal conditions.

2.4.5 Plutonium and Oxygen Migration

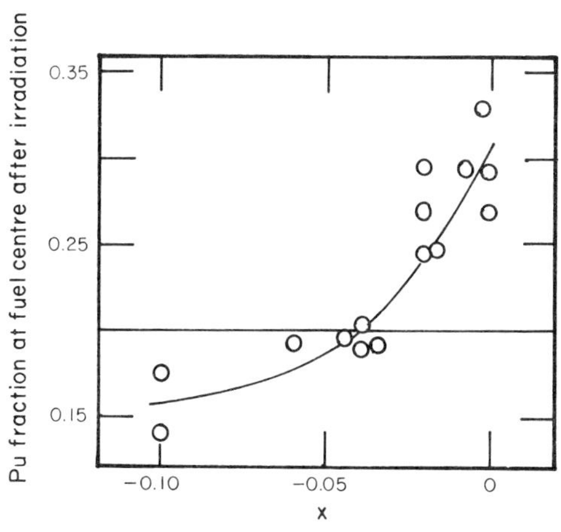

Fig. 2.14. Effect of Initial ratio of oxygen to metal on plutonium migration in fuel with initial composition $(U_{0.8}\ Pu_{0.2})O_{2+x}$

It is found that, even if the plutonium is initially distributed uniformly throughout the fuel, after only a short irradiation (less than 1% burnup) the relative concentrations of uranium and plutonium have changed, but the nature of the redistribution depends on x (Meyer, 1973). Figure 2.14 shows experimental values of plutonium concentration at the centre of the fuel after irradiation, indicating that for $x > -0.04$ the plutonium moves outwards. The plutonium in the outer, cooler, part of the fuel is hardly affected. A typical distribution of plutonium in irradiated fuel is shown in Fig. 2.15.

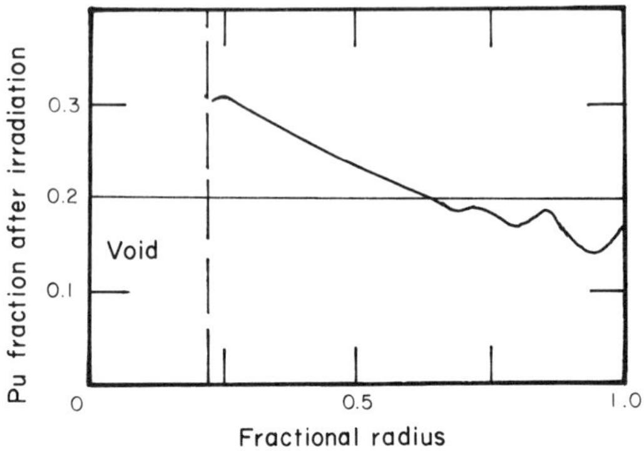

Fig. 2.15. Redistribution of plutonium in fuel with initial composition $(U_{0.8}\ Pu_{0.2})O_2$.

The mechanism for this redistribution seems to depend on the preferential evaporation and migration of different species. The stoichiometric ratio is important because if there is excess oxygen the mixture of vapours in equilibrium with the mixed oxide contains relatively more UO_3, while if oxygen is deficient it contains more PuO, and for $x < -0.03$ or -0.04 PuO predominates. Thus for more positive values of x it is uranium in the form of UO_3 which tends to evaporate from the hotter part of the fuel and condense in the cooler, while for more negative x plutonium in the form of PuO moves instead.

This is not the whole story, however, because it does not explain why the segregation is so limited in extent. It may be that diffusion of vapour through the dense columnar grain region is in fact severely restricted.

If the segregation of plutonium and uranium were more severe, it would have serious consequences. As the plutonium generates most of the power, its extensive migration to the centre would increase the fuel central temperature and possibly cause melting. It would also have the effect of delaying the Doppler feedback on reactivity (see section 2.3.2). For the limited segregation actually observed these effects are negligible.

The same mechanism also results in a tendency for the oxygen to migrate. For $x > -0.04$ the evaporation and recondensation of UO_3 results in oxygen moving outwards so that the fuel at the outside is nearer stoichiometric while that at the centre has even less oxygen. The build-up of oxygen in the outer parts of the fuel tends to increase the partial pressure of UO_3 there and so helps to oppose the outward movement of uranium.

2.4.6 Fission-product behaviour

TABLE 2.2 Fission Product Concentrations after 10% Burnup of Fast Reactor Fuel (Findlay, 1974)

Element		Concentration (atoms per initial heavy atom)
Krypton	Kr	.0020
Rubidium	Rb	.0017
Strontium	Sr	.0039
Yttrium	Y	.0021
Zirconium	Zr	.0202
Molybdenum	Mo	.0206
Technetium	Tc	.0058
Ruthenium	Ru	.0211
Rhodium	Rh	.0052
Palladium	Pd	.0137
Silver	Ag	.0016
Tellurium	Te	.0032
Iodine	I	.0016
Xenon	Xe	.0205
Caesium	Cs	.0187
Barium	Ba	.0065
Lanthanum	La	.0054
Cerium	Ce	.0129
Praseodymium	Pr	.0042
Neodymium	Nd	.0144
Promethium	Pm	.0017
Samarium	Sm	.0037

Only elements with concentrations greater than 10^{-3} listed.

Table 2.2 lists the fission products present in greatest abundance after 10% burnup of typical fast reactor fuel (30% plutonium with typical concentrations of the higher plutonium isotopes). The behaviour of a chemical system with so many components is obviously extremely complex and is certainly not understood in detail. The broad outlines are given below but the complexities are such that the actual behaviour in a particular fuel element with slightly different composition, cladding, temperature or irradiation history, may differ quite widely.

It is convenient (Powell, 1974) to divide the most abundant fission products into groups as follows. Elements in the same group behave roughly similarly.

(a) Inert Gases (Kr, Xe). These are mainly released from the fuel, but some are retained in solution or in small bubbles within the grains in the cooler parts of the fuel. See section 2.3.4.

(b) Alkali Metals (Rb, Cs). These are very volatile in elemental form and migrate to the cool periphery of the fuel. This is illustrated in Fig. 2.16, which shows the distribution of ^{137}Cs as determined by γ-spectroscopy. In some cases the isotopes which are daughters of inert gases, such as ^{133}Cs which is produced from ^{133}Xe which decays with a half-life of 5.3 d, and ^{87}Rb, produced from ^{87}Kr decaying with a half-life of 78 min, can appear in the gas plenum to which the precursors have migrated.

(c) Halogens (I). It is very difficult to determine what happens to the iodine because stable or long-lived isotopes are produced only in very

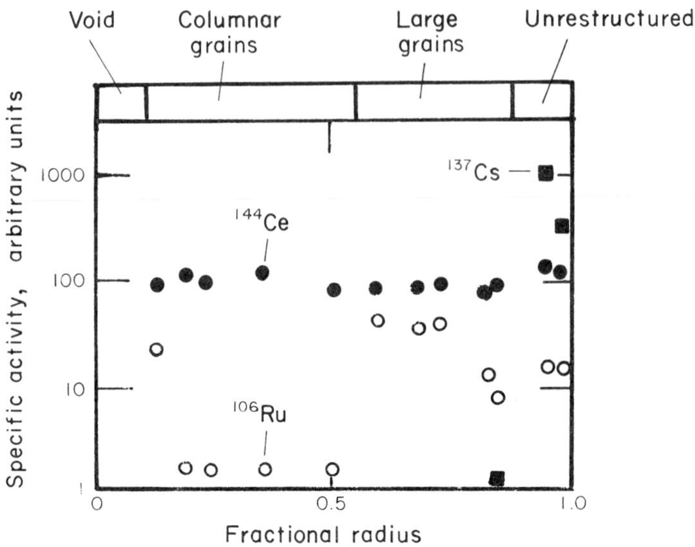

Fig. 2.16. Distribution of fission products in irradiated fuel.

small quantities (this applies to bromine as well), and ^{131}I decays with a half-life of 8.04 d. As it is volatile it seems to collect near the cladding and also at the ends of the fuel pins. It may be present as caesium iodide. Iodine may be involved in corrosion of the cladding.

(d) Metals forming refractory oxides (Sr, Y, Zr, Ba, La, Ce, Pr, Nd, Pm, Sm). By and large, having formed oxides these do not migrate and are found uniformly distributed through the fuel, as illustrated by the data for ^{144}Ce in Fig. 2.16. But again the isotopes which are daughters of volatile or gaseous precursors, such as ^{138}Ba and ^{140}Ba (from ^{138}Cs and ^{140}Cs respectively), are less uniform. ^{89}Sr (a daughter of ^{89}Rb which has a half-life of 15.4 min) migrates farther than ^{90}Sr (from ^{90}Rb, half-life 2.7 min).

(e) Metals which do not form oxides (Tc, Ru, Rh, Pd, Ag, Te). These are found as metallic inclusions, sometimes dispersed through the fuel and sometimes, especially if the central temperature is high, having migrated to the central void. There they form droplets of metal which are molten while the reactor is operating and are found as small ingots when the fuel is examined subsequently. Such ingots usually contain uranium and plutonium as well. Fig. 2.16 shows the loss of ruthenium from the hottest part of the fuel.

(f) Metals which may form oxides (Mo). The fate of the molybdenum depends on the oxygen potential of the fuel as explained in section 2.3.3. If it is low molybdenum is found in the metallic inclusions; if it is high it is found as MoO_2.

2.4.7 Corrosion of the Cladding

After the start of irradiation the inside surface of the cladding is soon made passive by forming a thin layer of oxide, mainly Cr_2O_3. As long as this layer stays intact the steel is protected from further corrosion. In some cases, however, it is disrupted in a few places and corrosion takes place, usually in a few isolated patches rather than uniformly.

Each breakdown of the oxide layer and subsequent corrosion seem to depend on the build-up of local concentrations of fission products, among which caesium and possibly tellurium or iodine appear to be most important. It may be that the caesium which migrates to the outside of the fuel is present there in a liquid phase which acts as an electrolyte and dissolves the Cr_2O_3.

Given access to the steel the fission products cause corrosion, which may be uniform and of only limited extent, but in some cases has been observed to be intergranular in nature and penetrating some 25 - 75 μm into steel at temperatures above 500°C. This can be a significant reduction in the effective thickness of the cladding, although few cases have been observed in which it has actually caused failure.

The corrosion mechanism is basically that of oxidation, which removes the chromium from the surfaces to which the fission products obtain access. The way in which it takes place is not clear: it may be a liquid-phase electrolytic process in which caesium ions act as intermediaries; or it may be a vapour-phase process involving the formation and breakdown of volatile ferrous iodide (Olander, 1976, p.186).

Whatever the details of the process it is certain that it is strongly dependent on the oxygen potential of the fuel and its severity is reduced when this is low. Thus there is an incentive to make sure that the oxygen concentration in the fuel stays below the stoichiometric value (see section 2.3.3). The difficulty is, as pointed out in section 2.4.5, that the oxygen tends to migrate to the outside so that the fuel there is always nearly stoichiometric.

CHAPTER 3

Engineering

3.1 INTRODUCTION

Having described the neutron physics of a fast reactor in Chapter 1 and the behaviour of the fuel elements in Chapter 2, in this Chapter we discuss the engineering of the core of a power-producing fast reactor. There are three main sections dealing with heat transfer, materials and structure.

Heat transfer comes first because the dimensions of the fuel elements and of the core are determined mainly by the demands of heat transfer. The fuel elements have to be of the right dimensions to transfer heat to the coolant at the required rate without getting too hot. The core has then to be large enough to allow enough coolant to flow through it to take the heat away.

Once the main dimensions are fixed the form of the core is determined by the properties of the materials of which it is made, and in particular by the way these properties are affected by neutron irradiation and exposure to sodium coolant. The structure of the core has then to be designed within these constraints to hold the fuel in place, to allow it to be changed when necessary, to distribute the coolant flow correctly, and to provide for the control rods.

3.2 CORE HEAT TRANSFER

3.2.1 Fuel Element Rating

The primary economic pressure is to maximise the power output from a reactor, because this gives the best return on the capital invested in the reactor plant and the inventory of fuel committed to the reactor and the reprocessing cycle. The heat generated, however, has to be removed from the reactor core, and the power is limited by heat-transfer considerations. The crucial limits are set by conduction of heat within the fuel elements and by the flow of coolant through the core.

The fuel elements are made for convenience in the form of long tubes of cladding, usually stainless steel, containing the fuel in the form of ceramic pellets or powder. If the power density in the fuel material due to fission is Q W m^{-3}, then ΔT_f, the temperature difference between the centre and the surface of a cylindrical fuel pellet, is given by

$$\Delta T_f = QR_f^2/4K_f \qquad 3.1$$

where R_f is the radius of the fuel and K_f is its thermal conductivity, assumed independent of temperature.

The heat transferred per unit length of the fuel element, known as the "linear heat rating", is $q = \pi R_f^2 Q$, and it follows from equation 3.1 that

$$q = 4\pi K_f \Delta T_f. \quad 3.2$$

The power density is limited by the maximum fuel temperature, which must not be too high, but as we saw in Chapter 2 it is easier to work in terms of a limiting value of q rather than a maximum fuel temperature, partly because K_f is not constant but a function of both temperature and time as the fuel is irradiated, and partly because it is very difficult to determine the temperature of the fuel in an operating reactor. For oxide fuel the maximum acceptable value of q is about 50 kW m^{-1} (see section 2.2.2).

If q is fixed, $Q \propto R_f^{-2}$, so the power density can in principle be increased indefinitely by reducing the radius of the fuel. A practical limit is set by the cost of manufacture which rises rapidly for very small fuel elements, and as a result the fuel radius cannot be less than about 2.5 mm. Thus the power density in the fuel is limited to about 2.5 GW m^{-3}.

3.2.2 Distribution of Power Density

While the reactor is operating the heat transferred from the fuel arises almost entirely from fission. Q is given by

$$Q \simeq \Delta E_f \sum_g \phi_g \Sigma_{fg}^{fuel}, \quad 3.3$$

where ΔE_f is the difference between the internal energy of the reactants and products of a fission event, assuming there is no temperature change and that the products are at rest. ΔE_f is about 200 MeV for ^{239}Pu and is only slightly different for other isotopes. This is about 3.2×10^{-11} J. Σ_{fg}^{fuel} is the group fission cross-section for the fuel material. It is not the same as Σ_{fg} in equation 1.3, which is an average cross-section for a region of the reactor including structure and coolant.

Equation 3.3 is not exact. Some of the energy is transferred by neutrons and radiation and appears in the structure and coolant and even in the shielding. Some is transferred when radioactive fission products decay and so appears after the fission has taken place. While the reactor is operating γ-heating in the structure, the outer parts of the breeder, and the shield, is important, and when it is shut down radioactive decay heating in the fuel is important. Nevertheless during operation some 97% of the energy appears promptly in the fuel.

The neutron spectrum is nearly the same throughout the core so Q is roughly proportional to the total flux, the distribution of which is shown in Figs. 1.10 and 1.11. The maximum value, Q_{max}, occurs at the centre of the core or at the inside of the outer enrichment zone. The average Q along the most highly rated fuel element is lower than Q_{max}, and the average Q for the whole core is lower still.

To take account of this "form factors" are defined. An axial form factor f_z is given by

$$f_z = \int_0^H Q(z)dz/HQ_{max} \qquad 3.4$$

where Q(z) is the value of Q at a distance z from the bottom of the most highly rated fuel element and H is the height of the core. The radial form factor, f_r, is then defined by

$$f_r f_z = \int_{core} Qdv/VQ_{max} \qquad 3.5$$

where the integral runs over the whole core and V is the core volume.

Both f_r and f_z are usually about 0.8, and the average power density over the whole core is about 0.64 of the maximum. The power density at the ends of the most highly rated element is about 0.4 Q_{max}, and at the extremities of the core it is as low as 0.2 Q_{max}.

If the peak linear rating is 50 kW m^{-1} therefore the average for the most highly rated element is about 40 kW m^{-1} and the average over the whole core is about 32 kW m^{-1}, while the least highly rated element has an average linear rating of about 18 kW m^{-1}. If a reactor is to produce 2500 MW of heat it requires a minimum of about 78 km of fuel elements in total, whatever their radius. If the radius is 2.5 mm the total volume of fuel is at least 1.5 m^3.

3.2.3 Coolant Flow

Two other important dimensions, the height of the core and the spacing between the fuel elements, are determined mainly, although not completely, by the coolant. The flow of coolant through the core is subject to limitations on temperature rise, pressure drop, and velocity, none of which can be too high.

If the height of the core is H, so that the power output from the highest rated fuel element is $q_{max} f_z H$, then

$$q_{max} f_z H = A v_{max} \rho c \Delta T_c, \qquad 3.6$$

where ρ and c are the coolant density and specific heat capacity, ΔT_c is the temperature rise as the coolant passes through the core, v_{max} is the mean velocity of the coolant associated with the highest rated element; A is the coolant flow area per fuel element and depends on the spacing of the fuel elements.

The temperature rise ΔT_c is fixed by need for adequate strength and resistance to creep in the cladding and structure (which determines the outlet temperature), and design of the steam plant and resistance to thermal shock (which determine the inlet temperature). It is usually about 170 K and there is little scope for increasing it significantly.

The maximum coolant velocity v_{max} cannot exceed about 10 m s^{-1}, partly because there is some evidence that erosion of steel cladding is significant at higher velocities and partly because vibration of fuel elements and structural components is much more difficult to control at higher velocities. This is however not usually so important a limitation as that on the pressure drop.

The coolant flow to less highly rated fuel elements is restricted, as explained below, so that the temperature rise is uniform across the core. The pressure drop is therefore determined by the most highly rated fuel. It may be thought of as being given by an expression of the form

$$\Delta P_c = C.\tfrac{1}{2}\rho v_{max}^2.(4H/D_H), \qquad 3.7$$

where D_H is the hydraulic diameter of the coolant channels (which depends on the separation of the fuel elements) and C is a constant with the nature of a friction factor. Equation 3.7 is useful as an illustration but in reality the situation is not so simple. The value of C depends on the details of the design of the coolant channels and especially on the nature of the fuel element supports.

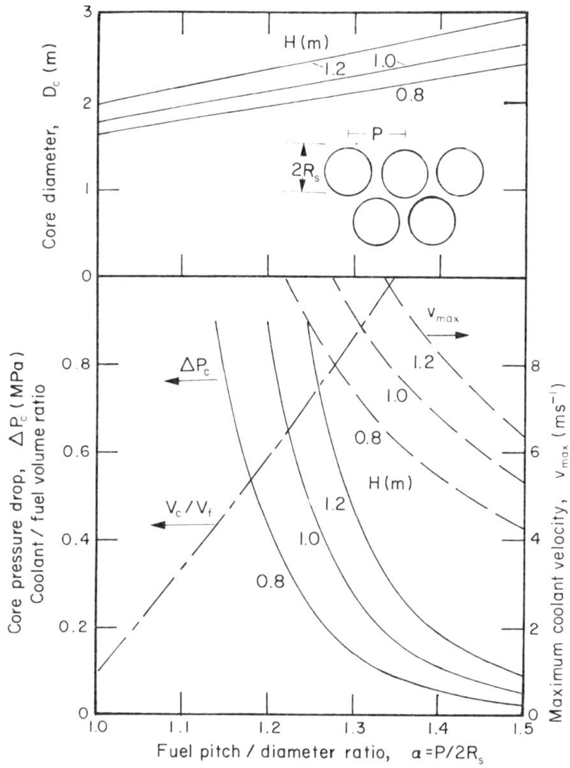

Fig. 3.1. Limits placed by heat transfer and coolant transport on core design: core diameter, pressure drop and composition, and coolant velocity vs. core height and fuel pitch-to-diameter ratio.

If the radius of the fuel elements is fixed then A in equation 3.6 and D_H in equation 3.7 are determined by the fuel element spacing. These two equations can then be used to determine v_{max} and ΔP_c as functions of H and the spacing. Typical results are shown in Fig. 3.1. The fuel elements are taken to be arranged in a triangular array as shown, the shape of which is determined by the ratio α of pitch to diameter, and the power is taken to be 2500 MW (heat).

The choices open to the designer can be seen. If for example ΔP_c is required to be no greater than 0.3 MPa, then choosing a core height of 1.2 m gives a reasonably compact core (Diameter/Height $\simeq$ 1.8) but the coolant volume fraction is very high. If the height is 0.8 m the coolant fraction is very much lower and the core diameter is now three times the height, which makes the critical mass large but keeps the sodium void coefficient of reactivity down. Higher values of ΔP_c are possible, but if they exceed about 700 kPa cavitation and noise generation may be important.

There are of course many other choices to be made as other parameters, such as the maximum linear rating and the core temperature rise, can be varied. The ultimate choice depends on the criterion used for optimisation. Economic criteria are important and safety (Chapter 5) has to be considered. But very often similarity to existing reactors which have been proved in operation is attractive because of the need to assure reliability.

3.2.4 Heat Transfer to the Coolant

Sodium is used as a coolant because it is a very good medium for heat transfer. As it is a metal its high thermal conductivity is due to the motion of the electrons in the conduction energy band, which are much more mobile than the positive ions. Being so light the electrons transport energy but little momentum so that the thermal conductivity is high but the viscosity is not. As a result the Prandtl number $Pr = \mu c/K$ (where μ is the viscosity, c the specific heat capacity and K the thermal conductivity), which for most fluids is of the order of one because energy and momentum are both transported by the diffusion of molecules, is very small for liquid metals. For sodium at 600°C, $Pr = 4.2 \times 10^{-3}$ and $K = 62.3$ W m^{-1} K^{-1}, while for water at 100°C, $Pr = 1.72$ and $K = 0.68$ W m^{-1} K^{-1}.

This does not mean, however, that the heat is transferred 100 times more readily to sodium than to water because in a flowing liquid the bulk motion as well as the diffusion of molecules or electrons is important. The higher the Reynolds number the smaller the advantage of a liquid metal over a non-metallic fluid.

The heat transfer coefficient h between a heated surface and a fluid is defined by $h = q_s/\Delta T$, where q_s is the local heat flux (W m^{-2}), and ΔT is the difference between the local temperature of the heated surface and the mixed mean fluid temperature. For fluid flowing in a channel h can be non-dimensionalised conveniently in the Nusselt number $Nu = hD_H/K$, where D_H is the hydraulic diameter of the channel. Dimensional analysis then shows that $Nu = Nu\,(Re, Pr)$, where the Reynolds number $Re = \rho v D_H/\mu$ and v is the mean fluid velocity.

For liquid metals, however, it is usually assumed that the heat transfer is independent of viscosity. This gives $Nu = Nu\,(Pe)$, where the Peclet number $Pe = RePr$ is independent of μ. Figure 3.2 shows the variation of Nu with Pe for flow in circular tubes.

As $Pe \to 0$, Nu tends to a constant value. If the liquid velocity were uniform across the tube and heat transfer were purely by conduction Nu would be 8. The effect of the variation of velocity across the tube is to reduce Nu, while turbulence increases Nu at high Pe. These effects can be calculated and, based on the work of Martinelli (1947), Lyon (1951) predicted

$$Nu = 7 + 0.025\,Pe^{0.8}. \qquad 3.8$$

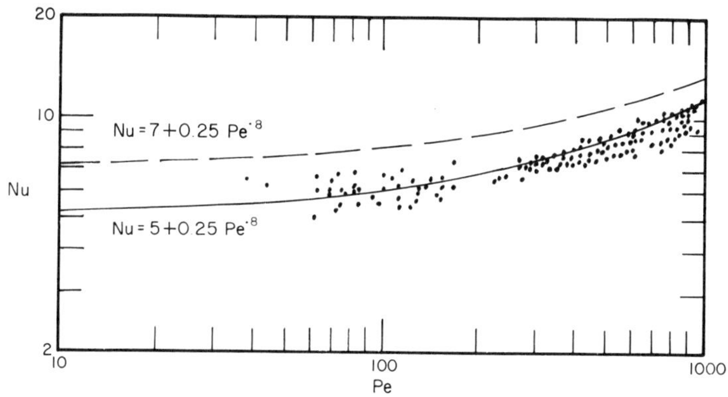

Fig. 3.2. Heat transfer to sodium in cylindrical tubes (Subbotin and others, 1963).

Figure 3.2 shows this to be on the high side of most of the data, and Subbotin and others (1963) suggest

$$Nu = 5 + 0.025\ Pe^{0.8} \qquad 3.9$$

as a better fit.

For flow at high Reynolds numbers parallel to a triangular array of cylindrical rods with uniform heat flux, Friedland and Bonilla (1961) propose

$$Nu = 7 + 3.8\alpha^{1.52} + 0.027\alpha^{0.27}\ Pe^{0.8} \qquad 3.10$$

for $\alpha > 1.35$, where α is the ratio of pitch to diameter. This equation is based on calculation, not experiment. Dwyer (1968) has performed calculations for smaller values of α, the results of which are shown in Fig. 3.3.

The experimental heat transfer data, especially for arrays of tubes, are in most cases very scattered and the experimental values of Nu usually lie below, and sometimes a factor of two below, the theoretical predictions. This is due in part to the difficulty of doing the experiments. If for example Nu = 8 for sodium flowing in a tube 10 mm in diameter, then $h = 5 \times 10^4\ W\ m^{-2}\ K^{-1}$. Even if the heat flux is as high as 100 $kW\ m^{-2}$ the temperature difference is only 2 K and cannot be measured accurately. In addition the temperature difference can be affected considerably by even very slight contamination of the heater surface. The presence of oxide films may explain why experimental values of Nu for sodium are frequently lower than those for mercury.

What is true for the experiment is also true for the reactor, and in many cases uncertainty in Nu is not important because the temperature difference is so small. If $q = 50\ kW\ m^{-1}$ and the outer radius of the cladding is 3 mm the heat flux is 5 $MW\ m^{-2}$. If the mean coolant velocity is 10 $m\ s^{-1}$ and $\alpha = 1.25$, Pe = 353. Figure 3.3 then gives Nu = 13, $h = 1.9 \times 10^5\ W\ m^{-2}\ K^{-1}$ and $\Delta T = 26$ K. This is the peak value at the most highly rated point in the core and it is clear that even a 20% error in h is not very important.

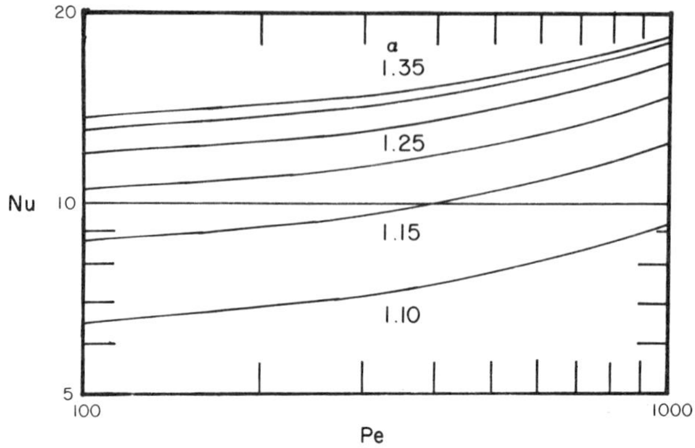

Fig. 3.3. Heat transfer to liquid metals flowing axially through rod bundles for low a (Dwyer, 1968).

3.2.5 Coolant and Cladding Temperature

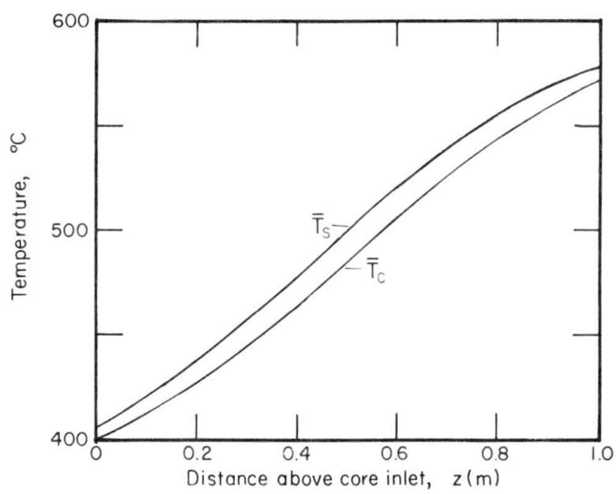

Fig. 3.4. Coolant and cladding temperatures through a typical core.

The mean coolant temperature $\overline{T}_c(z)$ at a height z above the bottom of the core, neglecting heat generated in the axial breeder, is

$$\overline{T}_c(z) = T_{ci} + \int_o^z q(z)dz/Av\rho c, \qquad 3.11$$

where T_{ci} is the coolant inlet temperature. The mean cladding temperature $\overline{T}_s(z)$ is

$$\overline{T}_s(z) = \overline{T}_c(z) + q(z)/2\pi R_s h, \qquad 3.12$$

where R_s is the outer radius of the cladding. The variation of $\overline{T}_s$ and $\overline{T}_c$ with z is shown in Fig. 3.4. $\overline{T}_s$ follows $\overline{T}_c$ closely and the maximum value of $\overline{T}_s$ occurs at the top of the core. This is unlike the case of gas or water-cooled reactors where h is much smaller and $\overline{T}_s$ reaches its maximum near the centre of the core.

The maximum cladding temperature is a very important limit on the design of the reactor. It is set principally by the creep strength which the cladding has to have as explained above (section 2.3.1), and in turn it sets the coolant outlet temperature and hence influences the thermodynamic efficiency of the plant. The maximum cladding temperature is, however, not the same as the maximum value of $\overline{T}_s$ because for a number of reasons there are, or may be, local variations which give locally high cladding temperatures or "hot spots".

3.2.6 "Hot Spots"

There are many possible causes of "hot spots", of which the most important are as follows.

The variation of coolant temperature across the coolant channel between three adjacent fuel elements is significant and is reflected as a variation of the cladding temperature round a fuel element. T_s is higher at a point near a neighbouring element and lower opposite a gap between elements. This is shown in Fig. 3.5. The variation is larger for close-packed arrays, and is also larger for fluids with low Prandtl numbers.

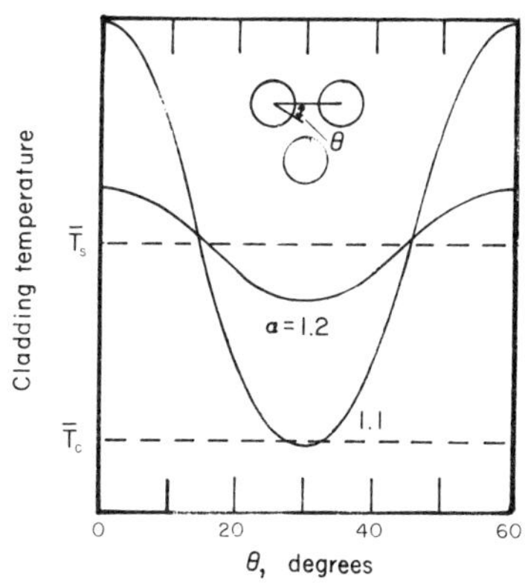

Fig. 3.5. Variation of cladding temperature round a fuel element for Pe ≃ 1000 (Dwyer, 1968).

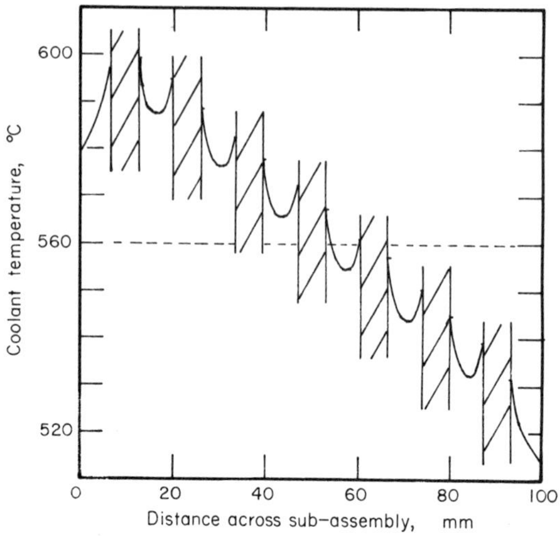

Fig. 3.6. Coolant temperature distribution at outlet from a 169-element subassembly, $\alpha = 1.3$, $R_S = 3$ mm, $\Delta T_C = 200$ K, $\overline{q} = 45$ kW m^{-1}, $q_{max}/\overline{q}$ (for subassembly) = 1.16 (Hoffmann and Weinberg, 1978).

This is because heat transfer within the coolant - i.e. from the intensely heated gap between two fuel elements to the centre of the subchannel between three fuel elements where most of the coolant flows - is strongly affected by the coolant velocity and is therefore relatively insensitive to Pr. Heat transfer across the boundary layer depends mainly on conduction and therefore varies strongly with Pr. If Pr is high boundary layer heat transfer is poor and the cladding surface is in effect partially insulated from the temperature variation in the coolant. The temperature variation around a fuel element in a water- or gas-cooled reactor is thus less severe than in a sodium-cooled reactor with the same value of q.

As explained below (section 3.4.1) fuel elements are normally grouped in "subassemblies" enclosed by hexagonal "wrappers". The coolant channels near the wrapper are often larger than those in the rest of the subassembly so that the edge fuel elements are overcooled. Coolant and cladding temperatures in the edge channels are lower than average, and this implies that the cladding temperatures in the centre of the subassembly are higher than average, giving another form of "hot spot".

Except at the centre of the core the power density varies across a subassembly and Q may be some 10% higher on one side than the other. As a result coolant and cladding temperatures on the hot side will be higher than the average, causing another "hot spot". The coolant temperature distribution across the outlet of a typical subassembly is shown in Fig. 3.6 which illustrates the effects of "power tilt", the overcooling of edge fuel elements, and the variation of temperatures across each coolant channel.

The fuel elements are held in position either by grids or wire wraps as explained in section 3.4.1. Either form of support modifies the coolant flow in places and although the effect is usually to increase the turbulence locally and hence increase h and decrease T_s there is a possibility of a region of poor cooling forming a "hot spot".

All these "hot spots" can be predicted, but there are others which occur at random because of variations of the dimensions and composition of various components. Such variations, even within manufacturing tolerance, can cause significant "hot spots". For example, Dwyer (1968) quotes the results of experiments indicating that if one fuel element is displaced by 0.3 mm from its correct position in the array there may be a decrease of the order of 10% in the local heat transfer coefficient. Variations in the thickness of the cladding, in the heat transfer between fuel and cladding, and in the fuel composition (affecting the local power density) can all cause variations in the heat flux and hence "hot spots".

Cumulatively the allowances that have to be made for all these effects are very significant. As a rough guide Aubert and others (1978) suggest

$$T_{s\,max} = T_{ci} + 15 + 1.5\Delta T_c \qquad 3.13$$

where $T_{s\,max}$ is the overall maximum cladding temperature (^oC), T_{ci} is the coolant temperature at inlet to the core and ΔT_c is the mean coolant temperature rise through the core. The additional 0.5 ΔT_c allows for all the "hot spots" which are proportional to coolant flow-rate (power tilt, edge channel overcooling, power density variation, etc.) while the 15 K is an estimate of the temperature difference between clad and coolant, allowing for variation of h round the fuel elements and due to variations in dimensions and compositions, at the top of the core where the temperatures are highest.

The important effect of the "hot spot" allowance in limiting the coolant outlet temperature can be illustrated as follows. $T_{s\,max}$ is typically limited to about 700°C to make sure that the cladding retains its integrity. If ΔT_c is 170 K, then equation 3.13 shows that the coolant inlet temperature is 430°C and the outlet 600°C. In spite of the excellent heat transfer properties of sodium, there has to be a margin of some 100 K to allow heat transfer under all conditions between cladding and coolant.

Heat transfer and fluid flow within the core are discussed at length by Tang, Coffield, and Markley (1978).

3.3 STRUCTURAL MATERIALS

3.3.1 Effects of Irradiation

The neutron irradiation to which the materials in the reactor are subject alters their properties in several ways. The most important effects on the structural materials are to increase hardness and decrease ductility, to enhance creep rates at low temperatures, and, most important, to reduce the density. These have to be taken into account in design of the reactor along with familiar effects such as thermal creep, fatigue and corrosion.

Irradiation affects the properties of non-fissile materials in two ways. Neutron scattering interactions displace atoms from their sites in the crystal lattice creating vacancies and interstitial atoms in equal numbers, and neutron absorption by (n,α) and (n,p) interactions creates atoms of helium and hydrogen and other transmutation products within the crystals. In many cases helium is the most important.

A useful way to characterise the extent of the irradiation a piece of material has received is to specify the average number of times an atom has been displaced from its lattice site. Each elastic scattering interaction imparts kinetic energy E_p to the target nucleus, where E_p is a random variable distributed uniformly (if the scattering is isotropic) in the range $0 - \mu E_n$, where E_n is the neutron energy and $\mu = 4A/(1 + A)^2$ where A is the atomic weight of the target. For iron $\mu = 0.069$.

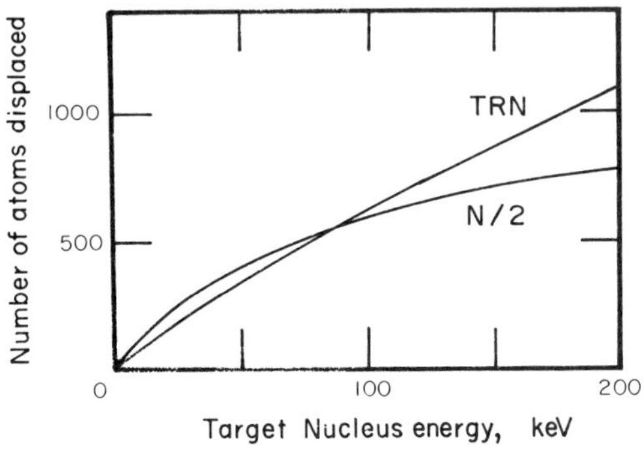

Fig. 3.7. Displacement of atoms in iron by nuclei which are targets of neutron scattering interactions, according to the "Half-Nelson" and "Torrens-Robinson-Norgett standard" models (Etherington and others, 1975).

It requires some 25 eV to displace an iron atom, so the target nucleus may well have enough energy to displace many atoms as it loses its kinetic energy and moves through the crystal. The relationship between E_p and n_d, the number of atoms displaced, is discussed by Etherington and others (1975), and values of $n_d(E_p)$ for two of the interaction models used most commonly are shown in Fig. 3.7. The number displaced does not increase indefinitely with E_p because at high energies inelastic scattering is increasingly important.

It is possible to define a "cross-section" for displacements for neutrons in group g, σ_{dg}, by

$$\sigma_{dg} \int_{E_g}^{E_{g-1}} \phi(E)dE = \int_{E_g}^{E_{g-1}} \sigma_e(E)n(E)\phi(E)dE, \qquad 3.14$$

where

$$n(E) = \int_o^{\mu E} n_d(E_p)dE_p. \qquad 3.15$$

$\sigma_e(E)$ is the elastic scattering cross-section, and $\phi(E)$ gives the variation of flux within the group from, for example, a fundamental mode calculation. Figure 3.8 shows values of σ_d for iron.

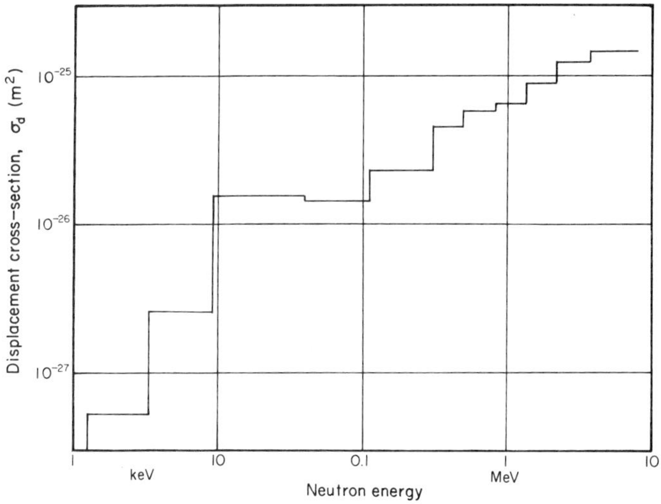

Fig. 3.8. Displacement cross-section for iron ("Half-Nelson" model, Etherington and others, 1975).

The total number of displacements per atom, D, is then given by

$$D = \int_o^T \sum_g \sigma_{dg}\phi_g(t)dt \qquad 3.16$$

where T is the length of time for which the irradiation continues.

It will be seen that σ_{dg} falls off rapidly at energies below about 0.1 MeV. For this reason an alternative way of characterising the extent of irradiation, often used because it is easier to calculate, is the fast neutron fluence, Φ_f, defined by

$$\Phi_f = \int_o^T \int_{0.1\ \text{MeV}}^{\infty} \phi(E,t)dEdt. \qquad 3.17$$

If Q is 2.5 GW m^{-3} in oxide fuel with a density of 10^4 kg m^{-3} containing about 20% PuO_2 and with a typical neutron spectrum such as that in Fig. 1.6 the total flux above 0.1 MeV is about 5 x 10^{19} m^{-2} s^{-1}. If the fuel is irradiated to 10% burnup Φ_f is about 1.4 x 10^{27} m^{-2} and D is about 70 displacements per atom in iron. This irradiation takes about 2.7 x 10^7 s or 10 months at full power. These conditions are typical of the centre of the core and represent the full extent of the

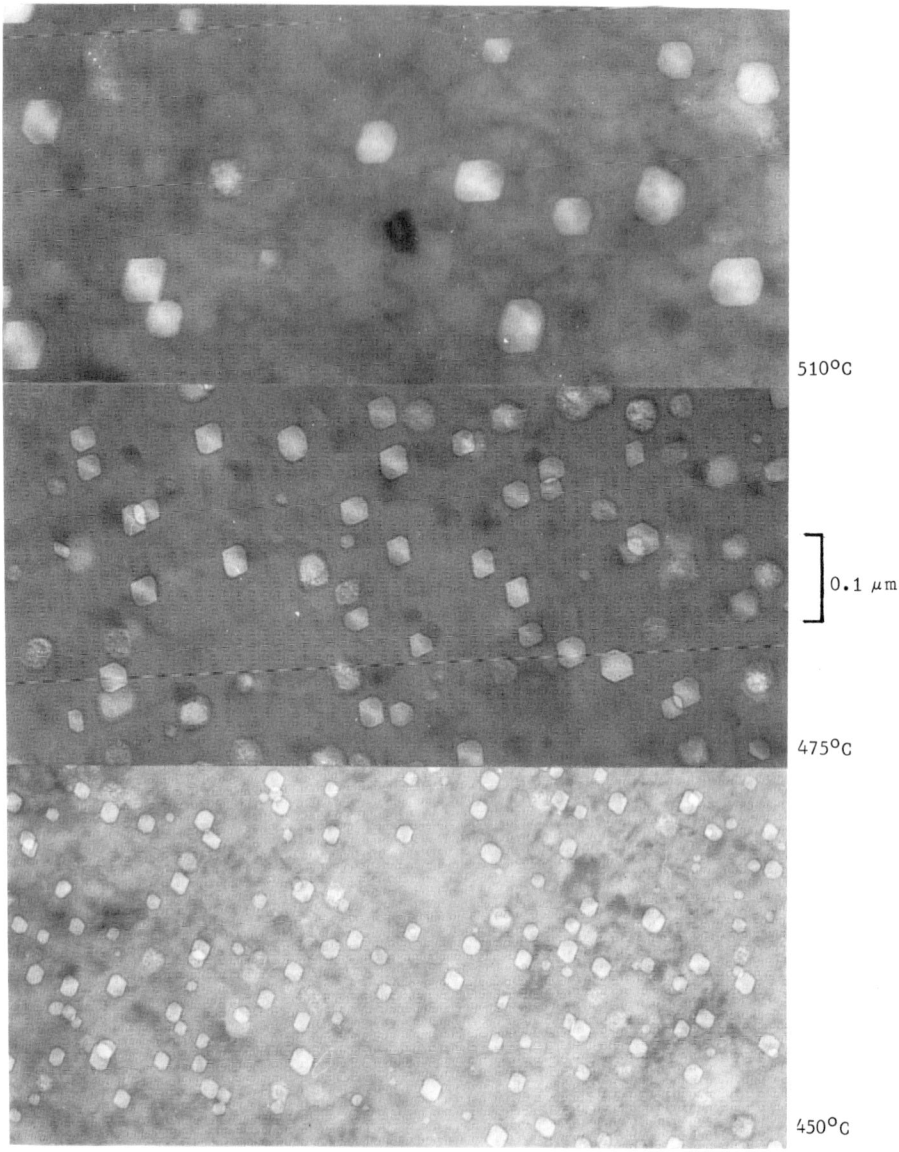

Fig. 3.9. Electron micrographs of 20% cold worked 316 stainless steel irradiated to about 44 displacements per atom ("Half-Nelson" model) at various temperatures.

irradiation suffered by structural material which is removed and replaced along with the fuel (i.e. cladding, subassembly wrappers, etc.).

Any material in the core which is not replaced along with the fuel, such as the control rod guides, suffers a higher fluence. It is therefore usual to design the core so that all its structure can be replaced if necessary. Items which cannot be removed such as the diagrid which supports the core are well below the core, where the fluence is lower typically by a factor of 10^4.

3.3.2 Density Reduction

The cloud of vacancies and interstitial atoms produced by a neutron scattering event diffuses through the crystal lattice under the influence of thermal agitation. If the motion were entirely at random the density of vacancies and interstitials would rise until production was balanced by recombination at sinks such as grain boundaries and dislocations where they would meet and annihilate each other. The motion is not entirely random, however. The stress field round a dislocation interacts with that round both interstitials and vacancies and tends to attract them to the dislocation, but the interaction with an interstitial is stronger. As a result the interstitials tend to cluster together at dislocations and other defects in the crystals, leaving an excess of vacancies which also tend to form clusters. The normal form of a cluster of vacancies is a flat mono-atomic layer which eventually collapses leaving an edge-dislocation ring, but if there is a nucleus, which may consist of a few atoms of an inert gas (helium), vacancies will migrate to it and form a three-dimensional void.

Figure 3.9 is a series of electron micrographs of irradiated AISI type 316 stainless steel showing typical polyhedral voids about 0.1 μm in diameter. As these voids are formed and grow the mean density of the material falls and it swells. Most metals swell in this way when irradiated but the rate and extent of swelling vary widely from one to another.

Temperature has an important effect on swelling as may be seen in Fig. 3.9. Figure 3.10 shows the ranges of values of void density and mean diameter typically observed. The greater size of the voids at higher temperature is probably due to the increased mobility of vacancies at higher temperatures, but the corresponding reduction in the number of voids is not entirely understood because the mechanism of nucleation is uncertain. It may be that at high temperature the helium atoms migrate to existing voids and are not available as nuclei of new ones. Alternatively nucleation may be connected with the "spikes" of displaced atoms due to a neutron scattering event, the damage due to which anneals out more readily at high temperature.

The result for many materials is that the swelling rate is high in a certain temperature range and low in others as shown in Fig. 3.10. For 316 stainless steel there is little swelling below 350°C. At high temperatures impurities may by very important and above 600°C the swelling is inhibited by the formation of very large voids on large grains of carbide. This is strongly dependent on the carbon concentration, and one way to reduce swelling is to reduce the amount of carbon in the material.

Swelling is also reduced if there are many dislocations because they tend to attract the vacancies, although weakly, and if there are enough of them the number of vacancies left to form voids is small. This is shown by the much reduced swelling in 20% cold-worked 316 stainless steel as compared with annealed material, especially below 500°C, as shown in Fig. 3.11.

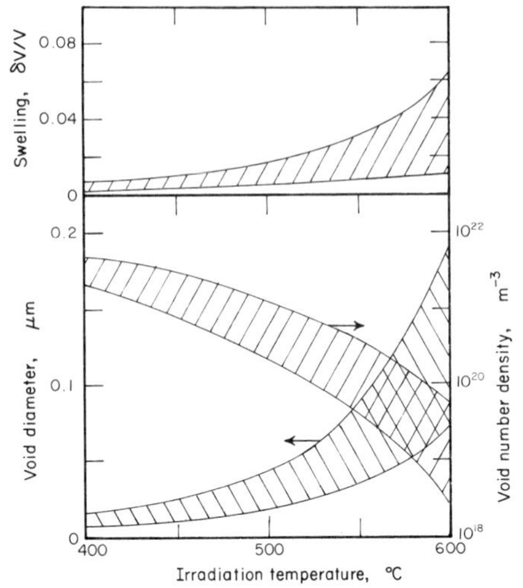

Fig. 3.10. Void diameter, number density, and swelling in cold-worked 316 stainless steel irradiated to 30-40 displacements per atom. (Bramman and others, 1978).

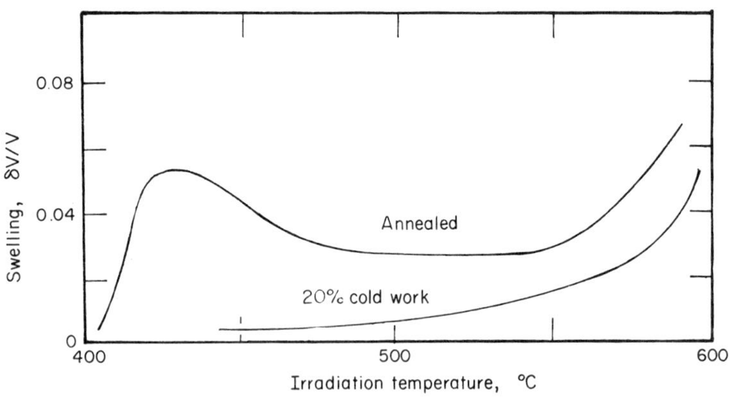

Fig. 3.11. Swelling of annealed and 20% cold-worked 316 stainless steel at 40 dpa (Bramman and others, 1978).

Finely-dispersed precipitates within the grains also tend to reduce swelling as they also attract both interstitials and vacancies and allow recombination. The swelling in the Nimonic PE16 alloy, in which the γ' phase is finely dispersed, is very low indeed. For this reason this material may be attractive for the structure of the core.

The amount of swelling increases with time and as it depends on the interaction between vacancies and nuclei the increase is non-linear. Initially swelling accelerates as irradiation proceeds, but at very high doses there is some evidence that it may saturate. Irradiation swelling is reviewed by Bagley, Bramman, and Cawthorne (1971), and more up-to-date information is given by Bramman and others (1978) and Brown and others (1979).

3.3.3 Irradiation Creep

The interstitial atoms produced by neutron scattering tend, as described above, to migrate to defects in the crystal. If the defect is an edge dislocation the accumulation of interstitials causes it to climb, and the material suffers strain. If it is subject to an overall stress, the stress fields of individual dislocations are modified according to their orientation with respect to the stress, so that the overall strain due to the climbing dislocations is in the direction of the stress and the material creeps. Another possible creep mechanism is that climb may free dislocations which have been pinned and immobilised by other defects, allowing them to glide.

Irradiation produces interstitials and therefore allows creep to take place at much lower temperatures than ordinary thermal creep. For example, substantial irradiation creep has been observed in 316 stainless steel at 280°C (Mosedale and Lewthwaite, 1974) whereas thermal creep is not significant below about 600°C.

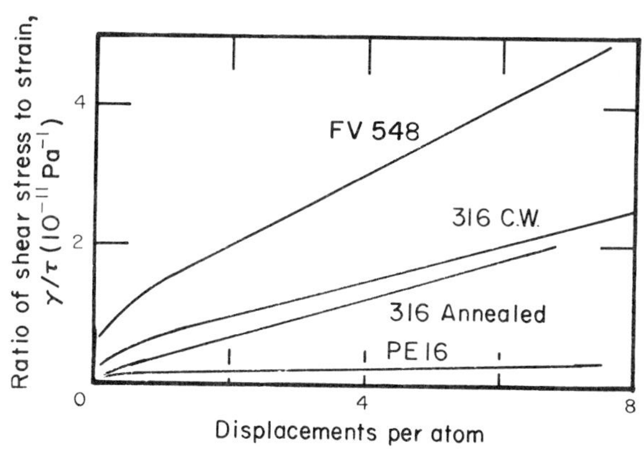

Fig. 3.12. Irradiation creep in various materials (Mosedale and Lewthwaite, 1974).

The creep strain is in most cases proportional to the stress and almost proportional to D. Figure 3.12 shows the ratio of creep shear strain γ to shear stress τ for various materials. These data can be represented by empirical formulae of the form

$$\gamma/\tau = A + BD. \tag{3.18}$$

TABLE 3.1 Irradiation Creep: Empirical Constants for Equation 3.18

Material	A (Pa^{-1})	B (Pa^{-1})
Annealed 316 stainless steel	0.2×10^{-11}	0.4×10^{-11}
20% CW 316 stainless steel	0.5	0.3
PE16 alloy	0.1	0.03
FV548 steel	0.9	0.5

Typical values of the constants are given in Table 3.1. The higher susceptibility to creep in cold-worked material is due to the greater density of dislocations. It is notable that PE16 alloy is less affected by irradiation than 316 stainless steel.

Irradiation creep is particularly important in reactor design because, as explained in section 3.4.2 below, it sometimes relieves the stresses caused by irradiation swelling.

3.3.4 Irradiation Embrittlement and Hardening

Irradiation both hardens materials (i.e. it raises the yield stress) and makes them more brittle (i.e. it reduces the elongation before failure). The ultimate stress usually changes relatively little, but there is a loss of work-hardening. Typical stress-strain curves are shown in Fig. 3.13.

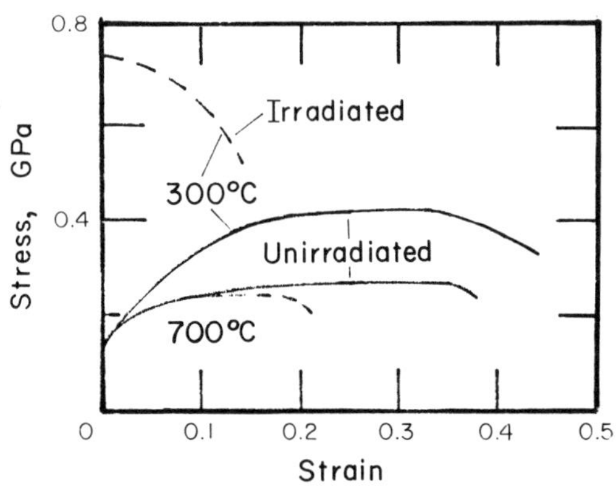

Fig. 3.13. Effect of irradiation to 15 dpa on stress-strain curves for 316 stainless steel (Bagley, Barnaby, and Fraser, 1973).

There are two mechanisms which cause these effects. At low test temperatures the defects caused by the irradiation damage reduce the mobility of dislocations and inhibit plastic strain so that the uniform elongation is very low. At higher test temperatures (500-600°C in 316 stainless steel), the material anneals, the defects are removed, and the properties of the unirradiated material tend to be recovered.

Above 700°C in 316 stainless steel, the second mechanism comes into play. As shown in Fig. 3.13 there is a loss of ductility which can outweigh the increase in ductility in the unirradiated material due to thermal effects. This is the result of helium generated by (n,α) reactions. At high temperatures it diffuses to the grain boundaries where it collects in the form of small bubbles. These cause a loss of cohesion between the grains, so that they can be torn apart.

Data on embrittlement and hardening of austenitic steels and a discussion of the mechanisms which cause them are given by Bagley, Barnaby, and Fraser (1973).

3.3.5 Corrosion and Wear

Sodium has the effect of removing nickel and chromium from the surface layers of austenitic stainless steel. The chromium forms sodium chromite, and the nickel is removed by dissolution in the sodium. Figure 3.14 shows the nickel and chromium concentrations near the surface of stainless steel exposed to flowing sodium. In this region the nickel concentration may be reduced to about 1% and chromium to 5 - 8%, and as a result a surface layer of ferrite is formed, some 5 μm thick (Nettley and others, 1967).

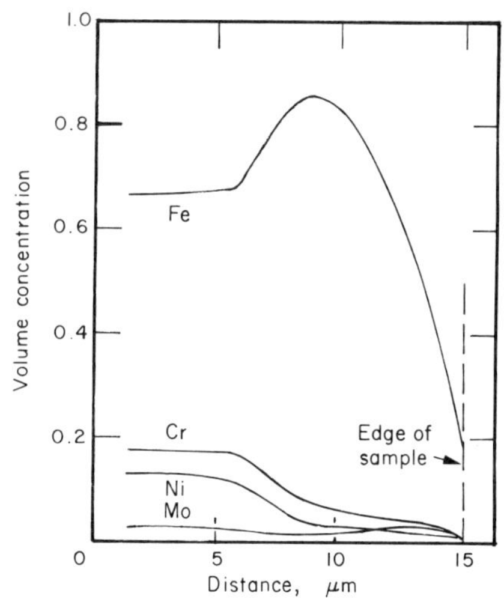

Fig. 3.14. Depletion of Cr and Ni from the surface of 316 stainless steel exposed to sodium at 650°C for 6 months (Nettley and others, 1967).

This ferrite is then dissolved at a rate which depends on the oxygen concentration in the sodium, and as it dissolves, the surface becomes rough. The greater the oxygen concentration, the greater the roughness. For 316 stainless steel, 10 parts per million of oxygen in the sodium gives a roughness of about 2 μm, while 25 ppm give about 6 μm, more or less independent of the velocity or Reynolds number of the flowing sodium (Thorley and Tyzack, 1973).

As both the formation of the sodium chromite and the dissolution of the ferrite depend on oxygen, the corrosion rate S is a function of the oxygen concentration P as well as the temperature T. Thorley and Tyzack (1973) correlate their experimental results by

$$\log_e S = 8.855 + 1.5 \log_e P - 9000/T \qquad 3.19$$

where T is in $^{\circ}$K, P in parts per million, and S in μm per year. S is independent of Reynolds number when Re is high and the surface roughness is comparable with the thickness of the laminar boundary layer, but falls off if Re $< 5 \times 10^4$, when the reaction rates are limited by diffusion through the boundary layer.

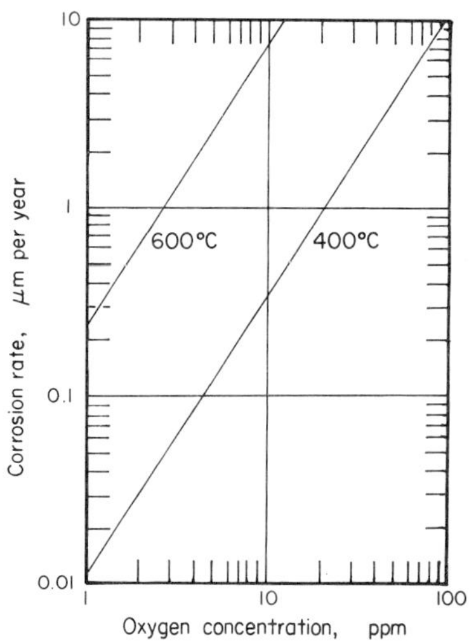

Fig. 3.15. Effect of oxygen concentration and temperature on corrosion of 316 stainless steel by sodium (Thorley and Tyzack, 1973).

Typical corrosion rates in 316 stainless steel as given by equation 3.19 are shown in Fig. 3.15. The corrosion rates of low-alloy steel, such as 2.25 Cr 1 Mo, and also in PE16 alloy with its high iron content are very similar to those of stainless steel because in all cases the controlling process is the dissolution of ferrite.

Corrosion, or more accurately wear, can be made worse by the relative motion of surfaces in contact, usually called "fretting". In a reducing environment such as sodium there is a tendency for surfaces in contact to become welded together at high spots. If the surfaces are forced to slide over each other, for example by some source of mechanical vibration, the welded high spots are sheared, and material may be transferred from one surface to the other.

Fretting damage is a danger in parts of the core subject to vibration due to turbulent flow of the coolant. Self-welding is a problem for items of equipment, such as fuel subassemblies, which have to be removed from the reactor from time to time. Hard wearing materials such as Stellite (an alloy consisting mainly of cobalt and chromium) are less prone to self-welding than stainless steels.

3.3.6 Carbon Transport

Depending on the activity of dissolved carbon, sodium can transfer carbon either to or from steels immersed in it. This process is known as carburization or decarburization depending on whether carbon is gained by the steel or lost from it. The rate of loss or gain of carbon depends on temperature because both the carbon activity in sodium and steel and the diffusion rates vary with temperature.

The carbon activity in austenitic steel is relatively low so it tends to be carburized even if the carbon concentration in the sodium is as low as 5 parts per million (Nettley and others, 1967). Carbide is precipitated in the surface layers of the steel, and this can reduce the ductility at low temperatures. Low alloy ferritic steels, however, such as 2.25 Cr 1 Mo with about 0.1% carbon, tend to be decarburized and lose strength as a result.

Carbon activity is higher at low temperature, so there is a tendency for decarburization to be particularly important if a sodium circuit incorporates ferritic steels in a low-temperature region such as a heat exchanger. If excessive carburization of austenitic steels in the high-temperature regions is to be avoided, careful control of the carbon activity in the sodium has to be maintained.

3.3.7 Thermal Stresses

The use of sodium as a coolant puts structural materials at risk of damage from thermal shock because temperature variations are transmitted very readily and result in thermal stresses which can cause fatigue or plastic flow. Temperature variations may be due to changes in reactor power, and in particular by emergency shut-down, which can cause the temperature of the coolant above the core to fall suddenly by 100 or 150 K.

The effect of these temperature changes can be shown by considering a sheet of structural material which for purposes of illustration we shall assume to be subject to no restraint except that it cannot bend. If the temperature varies through the thickness, and the strain is assumed to be uniform, it can be shown that the normal stress σ_s at the surface in a direction parallel to the surface is given by

$$\sigma_s = \alpha(\overline{T} - T_s)E/(1 - \nu) \tag{3.20}$$

where α is the linear coefficient of thermal expansion, E is Young's modulus, ν is Poisson's ratio, T_s is the surface temperature, and $\overline{T}$ is the average temperature of the sheet. (σ_s is taken to be positive in tension.) If the coolant temperature decreases suddenly by an amount T_o, and if the heat transfer coefficient is assumed to be constant, the variation of $\overline{T} - T_s$ can be found by the method described by Carslaw and Jaeger (1959).

A typical value for h for sodium flowing at 1 m s^{-1} over structural members with dimension of the order of 0.1 m is 14 kW m^{-2} K^{-1}. If the thermal conductivity of the material is 24 W m^{-1} K^{-1} (typical of stainless steel), the maximum value of $(\bar{T} - T_s)$ is about 0.49 T_o. With $\alpha = 19 \times 10^{-6}$ K^{-1}, $E = 2 \times 10^{11}$ Pa, and $\nu = 0.29$, the corresponding maximum stress is 260 MPa for $T_o = 100$ K. This can exceed the .2% proof stress in 316 stainless steel at 600°C, indicating that plastic deformation can be caused by the temperature changes due to emergency shut-down. In comparison, if the coolant were water the heat transfer coefficient under similar circumstances would be about 1 kW m^{-2} K^{-1}, and the peak stress would be about 60 MPa.

If there is a crack in the material any stress normal to the crack is concentrated by it, and in a region surrounding the tip exceeds the yield stress. Plastic flow occurs, and if the stress fluctuates each cycle may cause the crack to spread a little so that eventually the material fails. Austenitic steels are more susceptible to damage of this type than ferritic steels, and welded structures present special difficulties. The region of a weld may harbour residual stresses generated when it was made which may increase the effect of any applied stress, and inclusions or imperfections of the finish may form nuclei from which cracks can grow.

The need to ensure, with the high reliability essential for the reactor core structure, that members immersed in sodium will not fail in this way imposes significant constraints on the design. It is easier to ensure integrity if the temperature is lower, and this leads some designers (Anderson, 1978) to suggest lower core outlet temperatures to ease the design problems.

3.3.8 Choice of Structural Materials

It is perhaps as well at this point to summarise the advantages and disadvantages of the structural materials which may be used in the core. There are three main groups: austenitic steels, nickel alloys, and ferritic steels.

Austenitic steels of the AISI 304, 316 or 321 types are most widely used. They have relatively low yield strength but high ductility (which is particularly important for the fuel cladding). The creep rate can be reduced if the carbon content is reduced (as in the low-carbon 316 stainless steel) or if a stabilising additive such as niobium is included, as in FV548.

These steels resist corrosion adequately provided the oxygen content of the sodium is controlled. They suffer, however, from three disadvantages: irradiation embrittlement, susceptibility to damage by thermal shock, and, most important, irradiation swelling. Cold working to produce about 20% reduction of area reduces but does not eliminate the incidence of swelling (and also increases the yield stress), and irradiation creep mitigates the effects to some extent, but nevertheless the necessity to accommodate volume increases of 5 - 10% creates many difficulties in design.

High nickel alloys such as Nimonic PE16 are considerably stronger and are much more resistant to irradiation swelling. They have the disadvantage that the neutron capture cross-section in nickel is higher than that in iron, but this is balanced by the higher strength which allows a smaller volume of the material to be used. Alloys containing more nickel such as Nimonic 80A (about 75% nickel) corrode unacceptably fast in sodium.

Ferritic steels have rarely been used in reactor primary circuits because of the loss of strength they suffer on decarburization. This can be allowed for in design, and the effect is reduced if the carbon is stabilized by the addition of niobium or titanium. A steel such as 2.25 Cr 1 Mo then has the significant advantages of a higher thermal conductivity and lower thermal expansion coefficient than austenitic steels, both of which reduce thermal stresses, and greater resistance to crack propagation.

Table 3.2 gives typical compositions of some of these materials.

TABLE 3.2 Typical Compositions of Core Structural Materials (weight %)

	Austenitic Steels			Nimonics		Ferritic Steels
Element	316	318 (FV548)	321 (EN58B)	PE16	80A	2.25 Cr
Cr	17.3	16.3	18.0	16.6	19.4	2.3
Ni	13.8	11.6	9.0	43.1	76.0	0.2
Mo	2.4	1.3	0.5	3.4	-	1.0
Mn	1.8	1.1	0.5	0.1	0.1	0.5
Si	0.6	0.5	-	0.2	0.4	0.5
C	0.04	0.08	0.06	0.07	0.06	0.12
V	0.03	-	-	-	-	-
Nb	-	0.9	-	-	-	0.6
Al	-	-	-	1.2	1.2	-
Ti	-	-	0.5	1.2	2.3	0.4
Fe	64.0	68.2	71.4	34.1	0.5	94.4

3.4 CORE STRUCTURE

3.4.1 Fuel Subassemblies

The design of fuel elements is discussed in Chapter 2. Each element, or "fuel pin" as it is sometimes called, consists of a steel tube, 6 to 8 mm in diameter and usually some 2.5 to 3 m long. This allows for about 1 m of core fuel, 0.3 to 0.5 m of axial breeder fuel both above and below the core fuel, and a void or plenum up to 1 m long to contain the fission-product gases released from the fuel as explained in section 2.3.6. The plenum can be situated either below or above the core and axial breeders. If it is below the gas is cooler and therefore occupies a smaller volume, but there is the disadvantage that if the cladding should fail the gas as it is released it may perturb both reactivity and heat transfer as it is swept through the core by the coolant. A typical fuel element is shown in Fig. 2.6.

Fuel elements are usually arranged in a triangular array. This is better than a square array which, for the same ratio of coolant and fuel volumes, would require finer clearances between fuel elements and therefore greater variation of cladding temperature round each element.

As shown in sections 3.2.2 and 3.2.3 a reactor core may contain some 10^5 fuel elements, which obviously cannot be inserted and removed one by one. They are assembled into subassemblies each consisting of 200 - 300 fuel elements surrounded by a hexagonal tubular wall or "wrapper". Apart from facilitating handling this arrangement enables the coolant flow-rate to each part of the core to be regulated

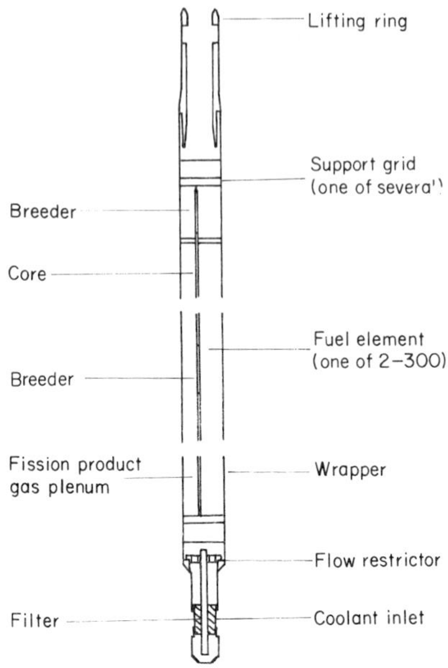

Fig. 3.16. Typical fuel subassembly.

so that the outlet temperature is approximately uniform across the core. This is done by means of adjustable flow restrictions or "gags" at the bottom of each subassembly.

The choice of subassembly size is a matter of compromise. Large subassemblies make for easier fuel handling, and lower manufacturing costs. Small subassemblies have the advantages of smaller transverse temperature variation (and therefore less severe hot spots), easier removal of decay heat when the irradiated fuel is removed from the reactor, and less serious consequences of an accident confined to one subassembly. (This point is explained more thoroughly in Chapter 5.)

Two methods of locating the fuel elements within the wrapper are in common use. In one a wire is wrapped helically round each element with a pitch of the order of 0.2 m. The diameter of the wire is equal to the required distance between elements so that each element is located with respect to each of its six nearest neighbours at two points in each pitch of the helix. Wire wraps tend to induce helical distortion of the fuel elements because they cause a slight asymmetry of the cladding temperature, and they introduce a considerable quantity of metal into the core. They have the advantage that the helical shape tends to promote mixing of the coolant between the various subchannels.

An alternative is to locate the fuel element by means of transverse grids every 0.1 m or so along their length. These can be made of spot-welded strips with pressed dimples which touch the elements, or they can be formed from solid material by spark erosion. They use less material and do not distort the elements as much as wire wraps, but they are expensive and have the disadvantage that the fine clearances between grids and elements may be prone to blockage by any foreign material present in the coolant. They also tend to have a high resistance to coolant flow which increases the pressure-drop through the core.

The subassembly may also contain shielding material, usually in the form of steel rods or a steel block, both above and below the fuel elements. This serves to protect the permanent structure on which the core rests, or equipment such as heat exchangers and the roof of the reactor vessel, from neutron and γ radiation from the core. There is also usually a coolant filter at the inlet end. Figure 3.16 shows a typical fuel subassembly.

Radial breeder fuel is usually contained in larger diameter fuel elements in sub-assemblies similar to those in the core.

3.4.2 Subassembly Bowing and Restraint

The subassembly wrappers become distorted due to the effects of thermal expansion, irradiation swelling and irradiation creep. The resulting movements have to be checked, otherwise they may cause unwanted reactivity changes, they may interfere with movement of the control rods, and they may make withdrawal of irradiated sub-assemblies difficult.

As explained in section 3.2.6 there may be a considerable temperature difference across a subassembly at the side of the core due to the radial variation of power density. The side of the wrapper nearer the core centre is hotter and expands more, causing the subassembly to become curved or "bowed" with the convex side facing the core centre.

Unless this motion is restrained, it has an effect on reactivity. If the sub-assemblies are mounted rigidly below the core bowing makes the core spread out, as shown in Fig. 1.17, and reduces the reactivity. If they were located above and below the core, however, bowing would tend to make the core more compact and add reactivity. This would be undesirable because it would be a source of positive feedback tending to make the reactor power unstable.

The effects of temperature bowing are in fact small. A temperature difference of ΔT across a subassembly of width w causes it to bow with a radius of curvature $w/\alpha\Delta T$ if it is not restrained. If the entire core section, of height H, of a subassembly held rigidly below the core (as in Fig. 1.17) is subject to the same ΔT so that its radius of curvature is uniform the centre is displaced outwards a distance $d \simeq H^2\alpha\Delta T/8w$. If $\Delta T = 10$ K, $\alpha = 2 \times 10^{-5}$ K^{-1}, H = 1 m and w = 0.14 m, $d \simeq 0.2$ mm. The resulting reactivity decrease is of the order of 10^{-4} which is only of minor importance.

The effect of irradiation swelling is similar in form but much larger. As swelling is an increasing function of fluence the inside of a peripheral subassembly, where the flux is higher, swells faster than the outside, again causing bowing with the convex side inward. If the difference in linear expansion between the two sides of the subassembly is $\Delta\epsilon$ the radius of curvature is $w/\Delta\epsilon$ and the displacement of the centre is $d \simeq H^2\Delta\epsilon/8w$ if $\Delta\epsilon$ is uniform.

$\Delta\epsilon$ varies along a subassembly in a complicated way because it depends on temperature as well as fluence. A mean value of 0.003, corresponding to a 1% difference in volumetric strain, is typical. This causes a displacement of about 3 mm at the level of the core centre, about 10 mm at the top of the core, and the displacement of the top of the subassembly could be 20 - 40 mm, depending on how long it is. This would create many problems, and in particular it would make it difficult to maintain the alignment of control rods with their operating machinery.

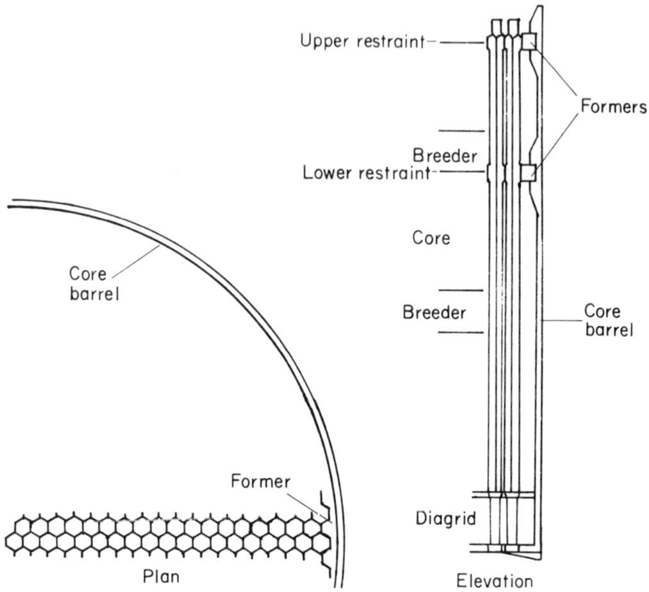

Fig. 3.17. Core restraint system.

The negative reactivity feedback resulting from supporting the subassemblies below the core and allowing them to bow outwards as the temperature increases is nevertheless an important advantage, and this method of support is used in the PFR for example (Frame and others, 1967). In most designs, however, radial movement is prevented by some sort of restraining or clamping system. Fig. 3.17 shows one such system, where radial movement of the outer periphery of the radial breeder is prevented at two restraint planes, both above the core. According to where the restraints are placed small movements of the fuel due both to thermal expansion and to irradiation swelling still occur, but any tendency to positive feedback can be eliminated.

A restraint system may be "active", meaning that after the subassemblies have been assembled the restraints are tightened so that the whole array is clamped together, or "passive", meaning that the restraints are fixed and prevent outward movements only after clearances between subassemblies have been taken up. The disadvantage of an "active" system is that the clamping mechanism has to be operated remotely and reliably under sodium.

To avoid self-welding between the wrappers they can be provided with hard pads at the points where they are in contact.

The loads between the restrained subassemblies vary with time and depend on the design of the restraint system, and predicting them is a complex task. The order of magnitude can be estimated fairly simply, however. If a simple cantilever of length L is subjected to a transverse force F at its free end, the displacement d is given by

$$d = FL^3/3EI, \qquad 3.21$$

where I is the second moment of area of the cross-section. For a hexagonal wrapper 0.14 m across flats made of sheet 3 mm thick, $I \simeq 4 \times 10^{-6}\ m^4$. If the length of the subassembly, L, is 3 m and the displacement is 0.01 m, typical as we have seen of bowing due to irradiation swelling, then with $E = 2 \times 10^{11}$ Pa we obtain $F \simeq 900$ N. If the restraint system is passive similar forces, of the order of 1 kN, may be needed to withdraw subassemblies.

3.4.3 Arrangement of the Core and Breeder

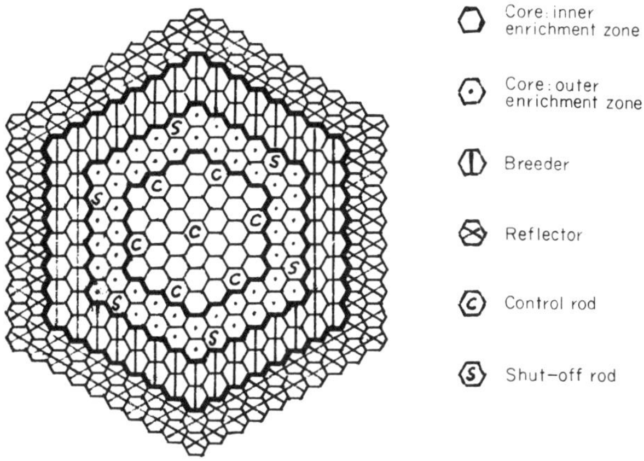

Fig. 3.18. Arrangement of subassemblies for a typical 250 MW (electrical) reactor.

The core and breeder are made up of an array of subassemblies of various types as shown in Fig. 3.18. The core subassemblies may contain fuel of several different enrichments arranged to give annular enrichment zones, as explained in section 1.3.3. The radial breeder usually consists of two or three rows of subassemblies consisting of fat fuel elements containing fertile material. The surrounding neutron shield may consist of subassemblies similar to those of the core and breeder and containing the shielding material which is usually steel, or it may be a separate structure.

Control rods occupy subassembly positions in the core region and are usually inserted into the core from above. They are operated by mechanisms which are situated on top of the reactor vessel in order to be available for maintenance.

Figure 3.18 shows a conventional core pattern with the fuel in two enrichment zones surrounded by a radial breeder. As explained in section 1.6.7, however, there are certain advantages in incorporating the breeder, or part of it, within the core, either in the form of one or more annular breeder zones, or by a more-or-less uniform distribution of breeder subassemblies within the core.

The subassemblies rest on a support structure, often known as a "diagrid", which also serves to distribute the coolant from the circulating pumps. A typical arrangement is shown in Fig. 3.19: each subassembly terminates in a hollow spike which is located in a hole in the diagrid, and coolant enters the subassembly through slots in the spike.

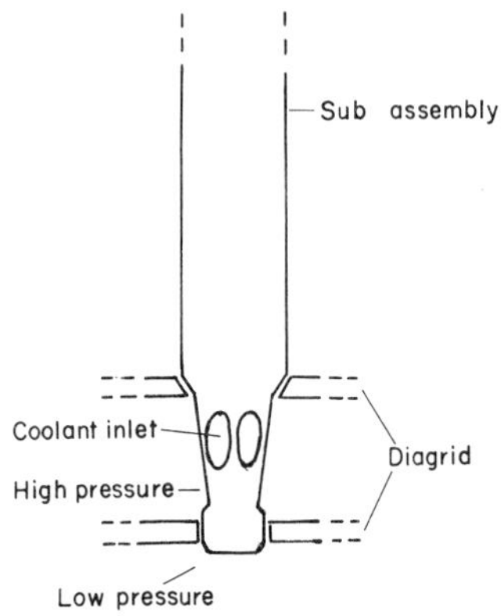

Fig. 3.19. Subassembly location and hold-down.

The pressure of the coolant in the diagrid is usually some 200-400 kPa above the hydrostatic pressure of the coolant outside, and this causes a substantial lifting force on a subassembly, which can exceed its weight. To prevent it floating away a "hold-down" mechanism is required. This can either be a rigid structure above the core which prevents upward movement, or an arrangement whereby the forces exerted by the coolant on the spike are approximately balanced. This can be done by arranging for the spike to pass right through the diagrid, so that its lower end is exposed to coolant at low pressure.

The wrappers contain high pressure coolant from the diagrid but the space between the wrappers is open at the top to low pressure coolant, so that there is a pressure difference across the wrappers of the order of 100 kPa. This induces stresses and in the core region, where the neutron flux is significant so that irradiation creep takes place, the flat faces of the hexagonal wrappers bulge outwards. If this brings neighbouring wrappers into contact self-welding may occur and withdrawal of irradiated subassemblies may be difficult. The size of the gap between wrappers, which is determined by the thickness of the hard pads where contact between sub-assemblies is allowed, must be great enough to prevent contact occurring anywhere else. Typically a gap of 6-8 mm is required.

3.5 DYNAMICS AND INSTRUMENTATION

3.5.1. Reactor Kinetics

The kinetic behaviour of the neutron flux can be described adequately in most cases by the "point kinetics" model, which can be expressed by

$$\dot{P} = (\rho - \beta)P/\tau + \sum_i R_i/\tau_i \qquad 3.22$$

with $$\dot{R}_i = \beta_i P/\tau - R_i/\tau_i \qquad 3.23$$

and $$\beta = \sum_i \beta_i \qquad 3.24$$

Here P(t) is the reactor power as a function of time, ρ is the excess reactivity defined by $\rho \equiv (k_{eff} - 1)/k_{eff}$, τ is the mean prompt neutron generation time, $R_i(t)$ is proportional to the number of neutron precursors of type i which decay with time constant τ_i, and β_i is the effective number of delayed neutrons produced from precursors of type i per neutron produced. β is the delayed neutron fraction.

The τ_i are only slightly different for fast or thermal fission or for different fissile isotopes and lie in the range from about 0.3 s to 80 s. The main difference between fast and thermal reactors is that the delayed neutron fraction is smaller for ^{239}Pu than for ^{235}U (.0021 as opposed to .0065) so that β is typically .004 for a Pu-fuelled fast reactor rather than about .007 for a thermal reactor. The exact value depends on the fuel composition and on the flux and importance spectra, and varies slightly with burnup.

τ is much smaller for a fast reactor than for a thermal reactor - about 10^{-6} or 10^{-7} s as opposed to about 10^{-4} s. For normal operation, however, it is quite unimportant and only under the most severe conditions can changes occur rapidly enough for τ to be significant.

The "point kinetics" model is valid provided the core is well-coupled, which means that changes in one part influence the neutron flux distribution everywhere essentially immediately. This is the case for conventional fast reactors in which the core is only a few mean free paths across. If the core consists of two or more regions each of which is nearly critical by itself and is only weakly coupled to the others the point kinetics model may be unsatisfactory.

The behaviour of the reactor at low power, when the effect of power on reactivity is negligible, is determined by the variation of ρ. The solutions of equations 3.22 - 3.24 for various cases are discussed in many texts. A thorough treatment of reactor kinetics, of which this section and the next are a brief summary, is given by Lewins (1978).

It is very useful to determine the response of the reactor to small sinusoidal perturbations of reactivity. If $\rho = \rho_o \exp(iwt)$, where $\rho_o \ll \beta$, then it can be shown from equations 3.22 - 3.24 that $P = P_o + p \exp(iwt)$ where $p \ll P$. p is given by

$$p = \rho_o P_o G_o(w) \qquad 3.25$$

where G_o is the zero-power transfer function given by

$$G_0^{-1} = iw(\tau + \sum_i \beta_i \tau_i/(1 + iw\tau_i)). \qquad 3.26$$

The amplitude and phase of G_0 are shown in Fig. 3.20 for a reactor with $\tau = 5 \times 10^{-7}$ s and $\beta = 0.004$.

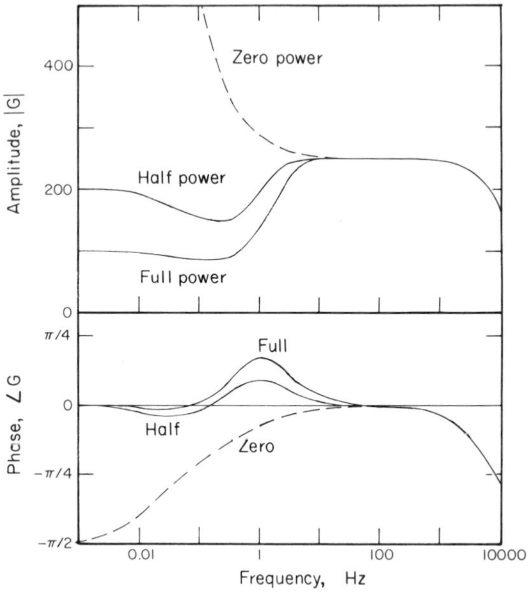

Fig. 3.20. Reactor transfer function at various lower levels.

3.5.2 Reactivity Feedback

When the power is high the behaviour of the reactor is determined by the effect of power on reactivity, which is illustrated by the simple closed-loop feedback diagram of Fig. 3.21. The reactivity ρ now has two components: the imposed reactivity ρ_1 which may be due to deliberate movement of the control rods, and the feedback reactivity ρ_f which depends on the power through a transfer function H(w). ρ_f is determined by the temperatures of various parts of the reactor and consists of the effects of thermal expansion of structure and coolant and of changes in resonance self-shielding (the Doppler effect). The temperatures in turn depend on the power P and on the primary coolant flow-rate and inlet temperature. The overall dependence of reactivity on power in the steady state is summed up by the power coefficient (see section 1.6.6).

When the reactor is not steady, however, account has to be taken of the fact that the temperatures, and therefore ρ_f, do not change immediately the power changes.

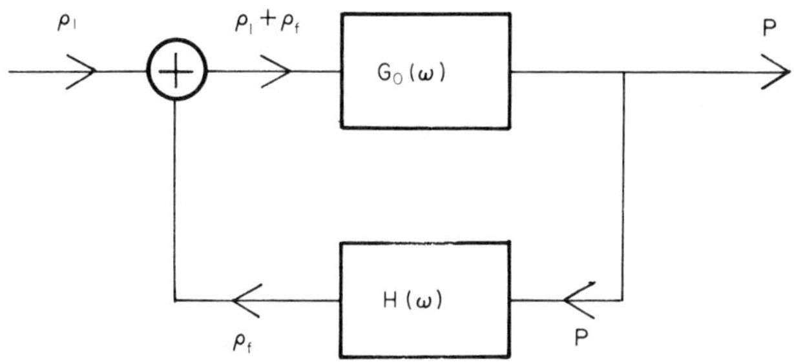

Fig. 3.21. Reactivity feedback.

The situation is very complicated because ρ_f depends on many different temperatures which respond to power changes in different ways, but as the most important effect is due to the fuel temperature a very simple illustration which shows the salient features is afforded by considering the average fuel temperature $\overline{T}_f$. This is given roughly by

$$\dot{\overline{T}}_f + (\overline{T}_f - \overline{T}_c)/\tau_f \; \alpha \; P, \tag{3.27}$$

where $\overline{T}_c$ is the average coolant temperature and τ_f is the thermal time-constant of the fuel elements. If the main resistance to heat transfer between fuel and coolant is in the fuel itself it can be shown that

$$\tau_f \simeq D^2/23\alpha, \tag{3.28}$$

where α is the thermal diffusivity of the fuel and D is the diameter of the fuel elements. For $\alpha = 10^{-6}$ m^2 s^{-1} (typical for oxide fuel) and D = 6 mm, $\tau_f \simeq 1.6$ s.

If ρ_f varies linearly with $\overline{T}_f$ equation 3.27 implies that ρ_f follows changes in P with a simple first-order lag having a time-constant of τ_f. This can be represented by

$$\rho_f = pH(w)\exp(iwt) \tag{3.29}$$

where $$H(w) = C/(1 + iw\tau_f). \tag{3.30}$$

Here C is the power coefficient of reactivity.

Since $\rho = \rho_1 + \rho_f$ we can arrange equations 3.25 and 3.29 to give p in terms of ρ_{1o} (where the imposed reactivity is assumed to be given by $\rho_1 = \rho_{1o}\exp\,(iwt)$). We obtain

$$p = \rho_{1o}P_oG(w) \tag{3.31}$$

where $$G^{-1} = G_o{}^{-1} - P_oH. \tag{3.32}$$

G is the transfer function at power P_o. The phase and amplitude of G are shown in Fig. 3.20 for full power and intermediate power, assuming that the power coefficient C is such that the reactivity change on going from zero to full power is -0.01.

It can be seen that there is no possibility of oscillatory instability for this simple form of feedback. This is because the phase of G_o is always between 0 and $-\pi/2$, while that of H is between $\pi/2$ and π if C is negative, so that there is never any possibility of $G^{-1}\rightarrow 0$. It must be remembered, however, that this is a very simple model of the feedback. In reality H is a much more complicated function and if for some range of w it is in the same quadrant of the complex plane as G_0^{-1} there is a possibility that for some values of w and P_0, $G^{-1} \rightarrow 0$ and the reactor is unstable. This can happen only if H has at least one component giving positive feedback of power to reactivity. As we have seen (sections 1.6 and 3.4.2) it is possible to design the core so that this does not happen.

3.5.3 Instrumentation

The reactor is controlled by moving the control rods, and the operation of the control system depends on information about the reactor power and the coolant flow-rates and temperatures.

Coolant temperature can be measured by thermocouples at the core outlet. The main difficulty is to ensure that a thermocouple measures the mean temperature correctly for the coolant temperature is not uniform. Coolant from the edge of a sub-assembly is cooler than the rest and unless the flow is mixed by some device to enhance the turbulence a thermocouple may be exposed to a stream of unrepresentative coolant, and moreover as the coolant flow-rate changes the flow pattern at the outlet may change possibly bringing coolant from a different unrepresentative part of the subassembly to the thermocouple.

Other difficulties are caused by changes in the power generated in a subassembly by the movement of nearby control rods, and by burnup of the fuel. It is thus usually necessary to measure the temperature at a number of positions at the top of the core and use an average for the control system. The alternative of measuring the temperature farther from the core where the coolant has become more thoroughly mixed is less satisfactory because the delay allows more of the structure to experience a temperature change before corrective action is taken.

Coolant flow-rate in a pipe can be measured conveniently by an electro-magnetic flowmeter, which makes use of the electro-motive force induced when a conductor (the sodium) moves through a magnetic field. This can be provided by permanent magnets or coils on opposite sides of the pipe so that the field is perpendicular to the flow.

The reactor power is determined by measuring the neutron flux at some convenient point. The flux is so high that it is usually quite adequate to measure it in the shielding surrounding the radial breeder or outside the reactor vessel.

The flux-measuring instruments are normally fission chambers or BF_3 chambers. The main difficulty is that neutron flux has to be measured over the range from full power (maybe 2500 MW) to the shutdown level of 100 mW or less - a range of more than 10^{10}. This cannot be done by any single instrument. If the flux is high it is possible to measure the overall ionisation current, but when it is low it is necessary to count individual pulses and determine the average count-rate.

Even with these two modes of operation it may not be possible to cover the entire range with a single instrument. It is frequently necessary to have two or more sets of fission chambers in positions of different sensitivity. At low power instruments close to the core are used, while at high power other instruments deeper in the shield are brought into operation.

The high gamma flux from the radioactive primary sodium has to be allowed for. At the highest powers it may be possible to ignore it in comparison with the neutron flux because the energy of a fission event is so much greater than that of a gamma from the sodium. But at lower powers, when the sodium activity (with a half-life of 15 hours) may correspond to earlier high-power operation, it is necessary to compensate for the ionisation caused by gammas or to discriminate against them if the pulse-counting mode is in use.

Several measuring stations round the core are necessary, partly for reasons of reliability and partly to allow for changes in the flux shape due to movement of control rods or irregularities in the pattern of loading new fuel into the core. As burnup proceeds the sensitivity of the instruments changes as the flux at the periphery of the core and in the breeder increases relative to that at the core centre.

CHAPTER 4

Coolant Circuits and Steam Plant

4.1 INTRODUCTION

This chapter describes the engineering of the remainder of the plant in a fast reactor electricity generating station, apart from the reactor core which is the subject of Chapter 3. It concentrates on the parts of the plant peculiar to a sodium-cooled fast reactor, which are the primary and secondary coolant circuits. The steam plant is very similar apart from the boilers to that in a conventional fossil-fuelled power station so it is mentioned only briefly here, and the electrical equipment which is entirely conventional is not dealt with at all.

Only brief mention is made of those parts of the plant which although they are specific to fast reactor power stations present no unusual engineering problems in design or construction. Such components are the containment building, the coolant circulating pumps, the fuel handling machinery and the control system.

In Chapter 3 we have emphasised the way in which the designer of the reactor core is constrained within very narrow limits by the properties of the materials he uses so that there is a marked similarity between all such cores. The same is not true for the coolant circuits. The use of sodium dictates the size of the heat transfer surfaces, but not the form of the circuits or the heat exchangers. Thus there are two very different approaches to the layout of the primary circuit, which lead to "loop" or "pool" reactors. In the same way many different steam generator designs are possible, employing U-tubes, straight tubes or helical tubes, with either once-through or recirculating flow on the steam side.

So far all fast power reactors have used a secondary liquid metal coolant circuit because the risk of having water and radioactive primary sodium in the same heat exchanger have been judged unacceptable. With increasing experience of designing and operating sodium-heated steam generators, and of preventing leaks in them, the situation may change in the future, as the capital cost would be reduced if the secondary sodium circuits could be dispensed with. But as our purpose is to describe existing and imminent systems we shall consider only designs incorporating secondary sodium circuits.

4.2 PRIMARY CIRCUIT

4.2.1 Pool or Loop Layout

The primary coolant receives heat in the reactor core, flows to a region where the neutron flux is low to transfer its heat to the secondary coolant in the intermediate heat exchangers, and then returns to the core. In fixing the layout of this primary coolant circuit two main choices have to be made: whether the heat exchangers and circulating pumps should be in separate vessels from the core or in the same one, and whether the pumps should be before or after the heat exchangers.

A "pool" reactor is one in which the entire primary circuit is contained within a single vessel, as shown in Fig. 4.1. The core is surrounded by a neutron shield and round this are placed the pumps and heat exchangers (usually at least three of each). In a "loop" reactor, in the other hand, as shown in Fig. 4.2, the core is contained in a small vessel with the main neutron shield outside. Hot coolant from the core passes through pipes to the heat exchangers and then back to the core vessel.

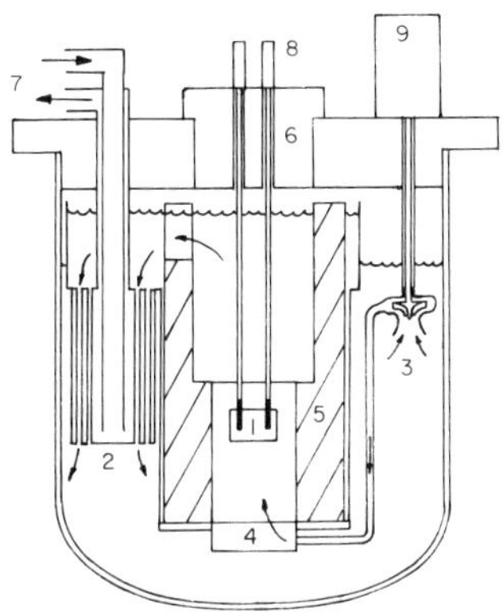

Fig. 4.1. "Pool" layout of the primary coolant circuit.

1. Core
2. Intermediate heat exchanger
3. Primary sodium pump
4. Diagrid
5. Neutron shield
6. Rotating plug
7. Secondary sodium pipes
8. Control rod mechanisms
9. Pump motor

The choice between the two schemes is affected by such considerations as the design and manufacture of the vessels, the design of the refuelling system, the operating conditions of the pumps, and ease of inspection and maintenance. That the choice is finely balanced is shown by the fact that reactors of both types have been built and are proposed for the future.

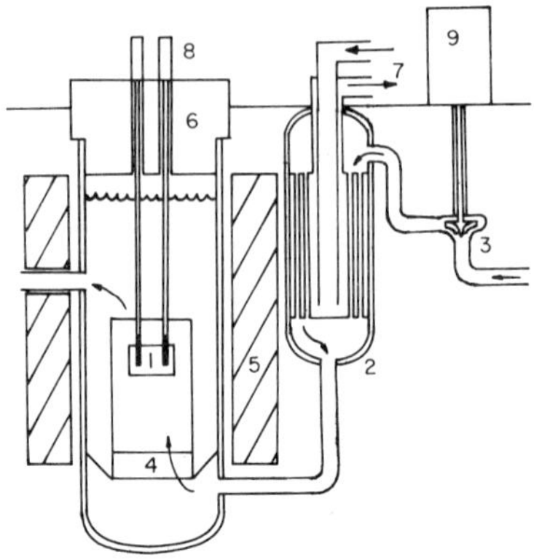

Fig. 4.2. "Loop" layout of the primary circuit.

1. Core
2. Intermediate heat exchanger
3. Primary sodium pump
4. Diagrid
5. Neutron shield
6. Rotating plug
7. Secondary sodium pipes
8. Control rod mechanisms
9. Pump motor

The main advantages of the pool layout are that the reactor vessel is very simple in form, the coolant pressure-drop is low, and the primary circuit can be arranged so that hot coolant never comes into contact with it. In a loop reactor parts of the pipework and of the vessel are in contact with hot coolant at temperatures at which thermal creep may be important, while other parts are in contact with cold coolant so that thermal stresses have to be allowed for. In addition there is the possibility of stress concentrations caused by the pipe branches on the sides of the vessel.

On the other hand a pool reactor vessel is so large that it has to be assembled on site, whereas a loop reactor vessel can be made in a factory where the quality of manufacture can be controlled more easily. The roof of a pool reactor vessel is much larger than that of a loop vessel, and if advantage is to be taken of the potential simplicity of the vessel itself the entire core, neutron shield, pumps and heat exchangers have to be suspended from it. Moreover part of the underside of the roof is exposed to the temperature of the hot coolant. As a result it is a complex and expensive structure.

A pool reactor has the advantage that there is room within the vessel for a temporary store for irradiated fuel, possibly in the form of a rotor as shown in Fig. 4.3. Fuel can be transferred from the core to the store without lifting it above the coolant so that no special provision has to be made for cooling it while in transit. It can be left in the store, immersed in coolant, until the fission-product power has decayed sufficiently to make handling easy. It can then be removed for reprocessing while fresh fuel can be kept in the store awaiting transfer to the core.

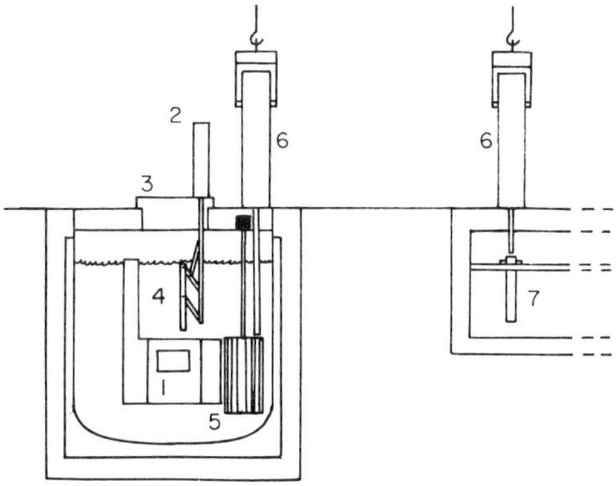

Fig. 4.3. Fuel handling system for a pool reactor with storage in the reactor vessel.

1. Core
2. In-vessel transfer machine
3. Single rotating plug
4. Pantograph transfer arm
5. Fuel storage rotor
6. Fuel transfer flask
7. Preparation for reprocessing

In a loop reactor vessel there is unlikely to be enough room for an irradiated fuel store. Irradiated fuel has to be removed from the vessel to a separate store by a machine which is capable of cooling it while it is in transit.

Both pool and loop reactors have pipework or structure operating at the temperature of the hot coolant. The difference is that in a loop reactor it is part of the primary coolant containment which is hot, and if it should fail radioactive primary coolant may be released. To offset this disadvantage, however, the loop system has the advantage that it may be possible to inspect the high-temperature pipework more easily since it is accessible from outside. It may even be possible to do maintenance work on one coolant loop by closing it off with valves without shutting down the whole reactor. Inspection and maintenance of the pipework and structure within a pool reactor vessel is difficult.

The advantages and disadvantages of loop and pool schemes are discussed in detail by Campbell (1973).

4.2.2 Pumps

Some early fast reactors used electromagnetic pumps to circulate the coolant, which have the advantage that no moving part penetrates the sodium containment. The sodium is pumped either by passing an electric current through it in the presence of a transverse magnetic field (a conduction pump), or by subjecting it to a moving magnetic field (an induction pump). The efficiency of electromagnetic pumps falls off with increasing size, however, and they cannot be used for circulating the main coolant flow in large reactors.

Mechanical pumps are now used exclusively. The problem of penetrating the sodium containment can be met by means of electric motors with totally-enclosed, "canned", rotors, but the usual method is to allow the shaft to pass through the containment above the sodium level. The penetration is thus exposed to the argon cover gas containing sodium vapour but not to liquid sodium. The seal is then purged continually with clean argon so that leakage of sodium or radioactive gases is prevented.

The pump is thus driven by a long vertical shaft, as shown in Fig. 4.1. At the bottom of the shaft is a bearing which has to operate immersed in sodium. This is usually of the hydrostatic type lubricated with sodium from the pump outlet (Eickhoff, Allen, and Boorman, 1967).

The pumps are driven by variable-speed motors and are also fitted with auxiliary or "pony" motors which are capable of turning the pump fast enough to maintain adequate flow of coolant to keep the fuel cool when the reactor is shut down. In the case of a power failure these can be energised from a standby power source such as a diesel generator to guarantee emergency cooling.

The pumps have to provide a large volume rate of flow at a relatively low pressure rise (typically about 15 $m^3 s^{-1}$ in total at 300 kPa for a 2500 MW (heat) pool reactor). For this single stage centrifugal impellers of conventional design are normally used. The main difficulty in design is to cope with sudden changes in temperature and to prevent cavitation, and in these respects loop and pool reactors pose different demands.

The coolant temperature has very little effect on cavitation, because even at 600°C the saturation pressure is only 7 kPa. The important factor is the pressure at the pump inlet (the "net positive suction head"). The pressure of the cover gas above the sodium in the reactor vessel is limited to some 1 - 200 kPa gauge to make it easy to prevent too much leakage of radioactive material and sodium vapour. In a pool reactor, the pump inlet pressure is this gas pressure plus the hydrostatic head of the sodium above the pump. This is normally adequate to prevent cavitation.

In a loop reactor, however, the suction pressure depends on whether the pump is before or after the heat exchanger (i.e. whether it is in the "hot leg" or the "cold leg"). If it is before the heat exchanger there is a loss of pressure due to the flow in the pipe from the reactor vessel to the pump. If it is after the heat exchanger there is an additional loss of pressure due to the flow through the heat exchanger and more pipe. It is not easy to compensate for these pressure drops by positioning the pump at a lower level to increase the hydrostatic head, and anyway this would require still longer pipes. Thus the conclusion is usually reached that cavitation is avoided more easily if the pump is placed in the "hot leg", before the heat exchanger. The relative advantages of "hot leg" and "cold leg" pumps are discussed by Campbell (1973).

If something goes wrong either the reactor or one of the secondary heat exchangers may have to be shut down suddenly. In a pool reactor the pumps draw from the large mass (of the order of 2000 tonnes or more) of cold coolant filling the vessel. If a secondary heat exchanger is shut down the temperature of this mass of coolant rises if the reactor is not also shut down immediately, but only slowly, and thermal stresses and differential expansion in the pumps are not significant.

In a loop reactor, however, the situation is quite different. If the reactor is shut down a "hot leg" pump is subject to a rapid fall in temperature, and if a secondary heat exchanger is shut down a "cold leg" pump is subject to a rapid rise. These temperature transients can be coped with, but they constitute another disadvantage of the loop design.

4.2.3 The Reactor Vessel

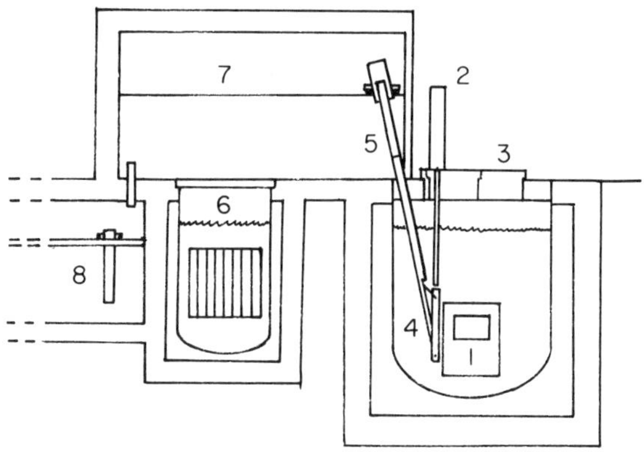

Fig. 4.4. Fuel handling system for a loop reactor with storage outside the reactor vessel.

1. Core
2. In-vessel transfer machine
3. Multiple eccentric rotating plugs
4. Fuel transfer position
5. Transfer hoist
6. Sodium-filled storage tank
7. Fuel transfer cell
8. Preparation for reprocessing

An advantage of the pool layout is that no penetrations through the side of the reactor vessel are needed. If in addition all the contents of the vessel - the core, shield, pumps, heat exchangers, etc. - are suspended from the roof, and if the vessel itself is suspended by its rim, it is stressed only by the weight of the sodium it contains. As the sodium in contact with the vessel is all at the core inlet temperature it experiences thermal stresses in the region of the sodium surface only. This means it can be of simple and light construction. A typical pool reactor vessel may be about 17 m in diameter and 14 m deep, containing about 2000 tonnes of primary sodium and made of stainless steel about 20 mm thick (Tattersall, Bell, and Emerson, 1967).

The vessel for a loop reactor may have inlet and outlet connections in the side walls, in which case it is more highly stressed and has to be of heavier construction. In some designs advantage is taken of this to support the vessel and the core from below. Some loop reactor designs, however, use a larger vessel so that there is room for the inlet and outlet pipes to pass through the roof and the vessel itself can be simple in form and lightly stressed like a pool reactor vessel. This gives the additional advantage that more space for fuel handling or storage within the vessel may be made available.

The main vessel is usually surrounded by a second "leak jacket" or "guard vessel", so that even if the main vessel should break the sodium level cannot fall below the top of the core and emergency cooling can be maintained.

Simplicity of the vessel has to be paid for with complexity of the roof. For a pool reactor this has to be a massive structure to support the contents of the vessel as shown in Fig. 4.1. It is usually in the form of a cellular steel girder filled with concrete or some other neutron shielding material. Part of the underside is exposed to cover gas at the core outlet temperature, so that some form of cooling may be needed.

Access through the roof to the fuel can be provided in a number of ways. Figure 4.4 shows a scheme involving a simple fuel handling machine which has to be positioned directly above a fuel subassembly to lift it. Such a machine has to be mounted on the inner of two eccentric rotating plugs, as indicated, so that it can be manoeuvred above any position in the core or breeder. A more complex fuel handling machine which can reach to the side to lift a fuel subassembly, as shown in Fig. 4.3, may need only a single rotating plug. Escape of cover gas round the plugs can be prevented by dip seals containing a liquid metal.

The control rod operating mechanisms are situated on the rotating plug or plugs. During refuelling these have to be disconnected from the control rods, which are left in the core to hold the reactor subcritical.

Within the vessel hot and cold coolant have to be separated by a barrier, sometimes called a "barrel" or "skirt", which surrounds the core and from the top of which the coolant flows to the intermediate heat exchangers. This barrier has to be insulated or protected in some way to prevent transfer of heat from hot primary coolant to cold.

4.2.4 Intermediate Heat Exchangers

The intermediate heat exchangers, in which heat is transferred from primary to secondary sodium coolant, are normally of a shell-and-tube design. Differential expansion of tubes and shell can be accommodated by means of expansion bellows or by bends in the tubes. To eliminate any possibility of radioactive primary coolant leaking into the secondary circuit the pressure of the secondary coolant in the heat exchanger has to be greater than that of the primary.

Because the coolants transfer heat so readily it is possible to keep the temperature difference between primary and secondary small and yet keep the intermediate heat exchangers reasonably compact. For example with tubes 20 mm in diameter and coolant velocities of about 5 m s^{-1} a Nusselt number of about 10 on both shell and tube sides is possible (see section 3.2.4), giving surface heat transfer coefficients of about 3×10^4 W m^{-2} K^{-1}. If the wall thickness is 1 mm this gives an overall heat transfer coefficient of about 10^4 W m^{-2} K^{-1}. The heat transfer rate Q, heat transfer area A, and the logarithmic mean temperature difference ΔT_m are related by

$$Q = UA\Delta T_m, \tag{4.1}$$

so that if Q = 2.5 GW and ΔT_m = 30 K an area of 8000 m^2 is needed for heat transfer. If the tubes are 8 m long and 20 mm in diameter some 16000 of them are needed, which could for example be arranged in 12 separate units each containing about 1300 tubes. Each unit would be a cylinder about 1.1 m in diameter, and all 12 could be incorporated in the vessel of a pool reactor. A loop reactor might have four separate loops each with a heat exchanger containing 4000 tubes in a diameter of about 2 m. The design of intermediate heat exchangers is discussed by Tang, Coffield and Markley (1978) p. 319.

In a pool reactor such as is shown in Fig. 4.1 the primary coolant is driven through the intermediate heat exchangers by the pressure difference due to the difference in levels between the hot coolant within the core skirt, and the cold coolant outside it.

In the event of an accident it might become impossible to reject heat from the secondary coolant and an alterantive means of removing the heat due to decay of fission products in the fuel would be needed. For this reason an auxiliary secondary coolant system is provided. In a pool reactor this may take the form of a separate auxiliary heat exchanger in the vessel, in which heat can be transferred to an emergency or "decay heat removal" cooling system. In a loop reactor there may be a separate auxiliary cooling loop in the primary circuit, or arrangements for emergency cooling of the secondary circuits.

4.2.5 Oxygen Control

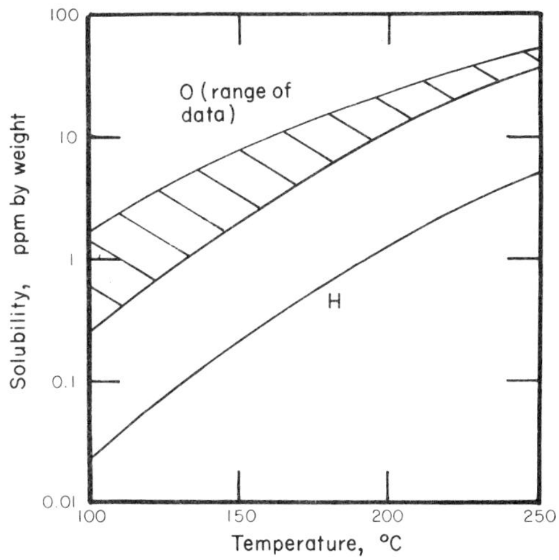

Fig. 4.5. Solubility of hydrogen (Whittingham, 1976) and oxygen (Claxton, 1976) in sodium.

As pointed out in section 3.3.5 oxygen dissolved in the sodium causes corrosion of stainless steel, and if the corrosion rate is to be limited the concentration of oxygen must not be too high. It is usually controlled at a level of about 10 parts per million by weight or less. On the other hand there is some evidence that a little oxide in the sodium acts as a lubricant preventing self-welding between steel surfaces in contact and facilitating the operation of immersed mechanisms.

The solubility of sodium monoxide Na_2O in sodium as a function of temperature is shown in Fig. 4.5. If the oxygen concentration is high oxide tends to be precipitated at low temperatures. This is another reason for controlling the oxygen concentration for precipitation of oxide in the cold parts of the circuit could block narrow passages such as coolant monitoring pipes.

Fortunately the low solubility of oxide at low temperature also affords means of controlling, and of monitoring, the oxygen concentration. It can be controlled by passing the sodium through a "cold trap". This is a device in which the sodium is cooled to a temperature below 160°C or so and then passed through a bed of stainless steel mesh or rings in which oxide is precipitated. In this way the oxygen concentration can be reduced to about 10 parts per million and oxygen which enters the circuit as an impurity in the argon cover gas or as moisture on new fuel can be removed.

To obtain lower oxygen concentrations a "hot trap" may be needed. In this the sodium is heated to 600 or 700°C and passed over zirconium, which has a greater affinity than sodium for oxygen. Hot traps and cold traps cannot be used together because oxygen would be transported from the cold traps to the hot traps. Hot traps can reduce the oxygen concentration to 1 - 2 parts per million.

4.2.6. Hydrogen Behaviour

Hydrogen is formed in the steam generators as the water oxidises the steel tubes. It diffuses readily through the tubes and enters the secondary sodium where it forms sodium hydride, NaH. The solubility of sodium hydride in sodium is also shown in Fig. 4.5. It behaves similarly to oxide, and is precipitated in cold traps in the same way. As there is a continuous source of hydrogen cold traps in the secondary circuits have to be operated continually to prevent precipitation of hydride in the coldest parts of the steam generators. Hydrogen can diffuse equally readily through the tubes of the intermediate heat exchangers into the primary sodium. If it is not trapped in the secondary circuits it appears in the primary cold traps.

Tritium, ^3H, is generated in the reactor core in three ways. It is formed directly in fission by the rare ternary fission events; by the neutron capture reaction $^6Li\ (n,\alpha)^3H$ in lithium present as an impurity in the sodium; and by the capture reaction $^{10}B\ (n,2\alpha)^3H$ in boron in the control rods. The total production rate is small, of the order of 100 Curies per day (2.4×10^{16} atoms per second) in a 2500 MW (heat) reactor.

It diffuses readily through the tubes of the intermediate heat exchangers and tends to be precipitated in the secondary cold traps because these are usually in continuous operation. This may cause difficulty in disposing of the cold trap packing when it is full of hydride because of the radiological effect of the tritium. It is radioactive with a half-life of 12.4 years, emitting a weak β.

4.2.7 Impurity Monitoring

Low temperature precipitation can be used to monitor oxygen or hydrogen concentration by means of a "plugging meter". The sodium is passed through a cooled orifice or passage and the flow-rate or pressure-drop is measured. When precipitation starts the passage tends to become "plugged" and the flow-rate falls or pressure-drop rises, depending on the design of the instrument. The temperature at which this occurs is called the "plugging temperature". It is related, but not equal, to the saturation temperature of oxide or hydride and is therefore a measure of concentration.

A problem arises, of course, when both oxygen and hydrogen are present, for the plugging temperature is a function of both oxide and hydride concentration. There is some evidence (Smith, Simm, and Hughes, 1979) that the two species are precipitated at different rates so that their effects can sometimes be separated by observing the dynamic behaviour of the cold trap. For operational requirements, however, the confusion is not too important because there is usually little oxide in the secondary sodium and little oxide or hydride in the primary.

An alternative approach is to measure oxygen concentrations by an electrolytic method (Hans and Dumm, 1977). The electrolyte is a ceramic (thoria doped with yttria) which separates the sodium forming one electrode from an air reference electrode. The potential generated depends on the oxygen concentration in the sodium. The main problem is that the ceramic electrolyte is very brittle and susceptible to thermal shock, so the temperature of the sodium has to be controlled very carefully.

Hydrogen concentration can be measured by utilising its ability to diffuse through a nickel membrane into a carrier gas such as argon (Hans and Dumm, 1977). The concentration of hydrogen in the argon can then be measured by a katharometer, which depends for its operation on the marked effect of the hydrogen on the thermal conductivity of the mixture. Alternatively the membrane can be evacuated by a vacuum pump of known pumping speed. The hydrogen pressure upstream of the pump, which can be measured by an ionisation gauge, depends on the rate of diffusion through the membrane, and therefore on the hydrogen concentration in the sodium.

4.3 STEAM PLANT

4.3.1 Steam Generator Design

The secondary sodium gives up its heat to raise steam in steam generators which are normally shell and tube heat exchangers with the water or steam in the tubes. These have to be larger than the intermediate heat exchangers because of the poorer heat transfer on the steam side. They also differ in that they are stressed by the high-pressure steam as well as by thermal expansion. The overriding concern in design and operation is to prevent leaks, because of the consequences of the chemical reaction between water and sodium.

At present most fast reactor power plants have recirculating Lamont-type steam generators with separate evaporator and superheater sections. Reheat can be accommodated by means of a reheater in parallel with the superheater. The flow of hot secondary sodium is divided between superheater and reheater and then recombined before flowing to the evaporator.

Austenitic stainless steel cannot be used in evaporators because of the risk of chloride stress corrosion. Although the chloride concentration in the feedwater can be controlled by ion-exchange units there is a danger of an accidental increase, particularly if the condenser is cooled with sea-water. The evaporator tubes can be made of ferritic steel, such as 2.25 Cr 1 Mo with about 0.4% niobium added to stabilise the carbon, or a steel with a higher chromium content such as 9 Cr 1 Mo which resists decarburisation. The superheater and reheater of a recirculating steam generator can be made either of austenitic steel (provided they can be kept free from droplets of water from the steam drum) or of a ferritic steel.

The most important consideration in determining the form of the steam generator is the nature of the joints between the tubes and the tube plate. They are exposed to sodium or sodium vapour, and the frequency of even very small leaks from the steam side must be very low.

If the tubes are in the form of a U, both ends can be positioned above the level of the sodium in the shell and protected from it by a layer of argon. This helps to reduce the thermal stresses at the joints, and accommodates relative thermal expansion of tubes and shell. Other forms such as the "hockey-stick" design or designs with helical tubes have bends in the tubes to allow for thermal expansion. If straight tubes are used there has to be a bellows in the shell. Various possible forms are shown in Fig. 4.6.

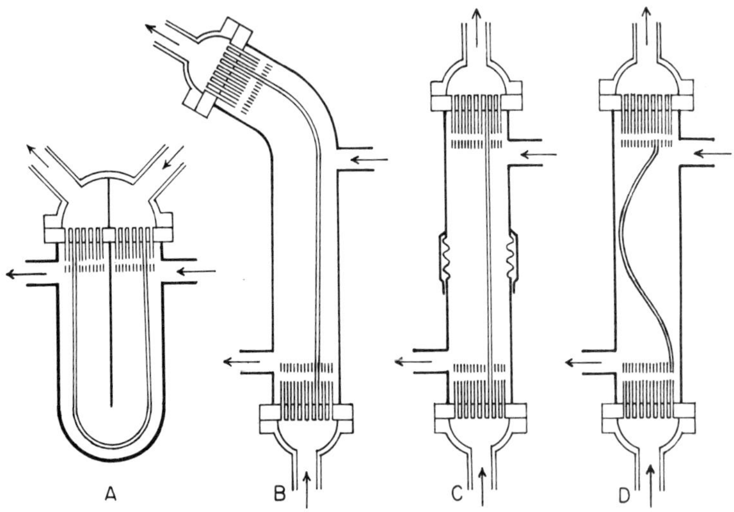

Fig. 4.6. Steam generator configurations.

A. U-tube
B. "Hockey stick"
C. Straight tube
D. Helical tube

The tubes can either be welded to a tube plate, as shown in Fig. 4.7, or they can pass through it to a manifold, with the gaps between tubes and tube plate sealed by "thermal sleeves" as shown in Fig. 4.8. If they are welded, the welds have to be of the highest quality, and the design has to be such as to make good welding possible.

The thermal sleeve design has the advantage that the joints between the sleeves and the tubes, and those between sleeves and tube plate, are not stressed by the steam pressure. A leak is not particularly serious, and the joints can therefore be brazed. The disadvantage is that in a large heat exchanger the manifold, which may have to connect many hundreds of tubes, is very complicated and expensive.

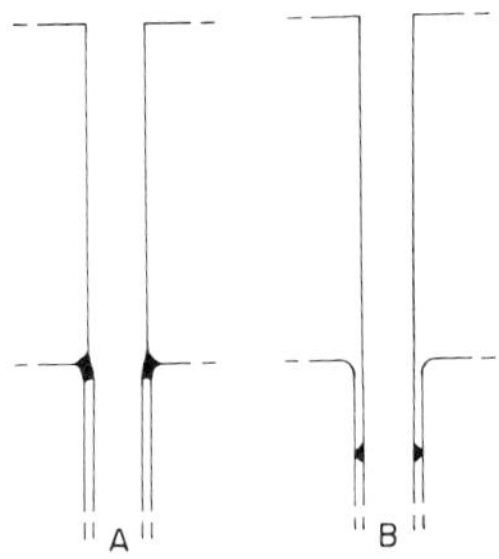

Fig. 4.7. Steam generator tube-to-tubeplate welds.

A. Fillet weld to inset tube
B. Butt weld to preformed spigot

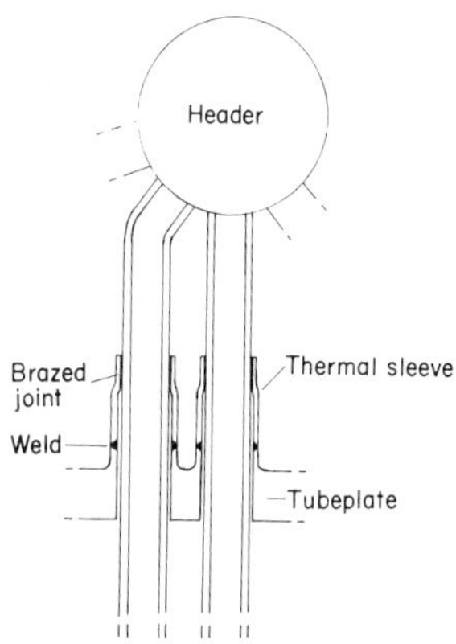

Fig. 4.8. Steam generator thermal sleeves.

The problems of steam leaks can be avoided almost completely by the use of double-walled heat exchangers with some means of detecting leakage of either steam or sodium into the space between the walls. The disadvantage is that the heat transfer coefficient is bound to be much worse than in a single-walled heat exchanger, and the cost will be higher.

If a single-walled design is chosen the risk of leaks has to be accepted. As explained in section 4.3.4 below a leak is not dangerous, but it does put the heat exchanger in which it occurs out of action until it is repaired. This drawback can be reduced by having a large number of separate small heat exchangers or heat exchanger modules, any one of which can be shut down for maintenance without reducing the power output of the whole plant by much. On the other hand it is cheaper to build a small number of large heat exchangers. In the end a choice has to be made between better plant availability and lower capital cost.

Useful discussions of steam generator design are given by Hayden (1976) and Lillie (1978) and of the performance of tube-to-tubeplate welds by Broomfield and Smedley (1979).

4.3.2 Steam Generator Heat Transfer

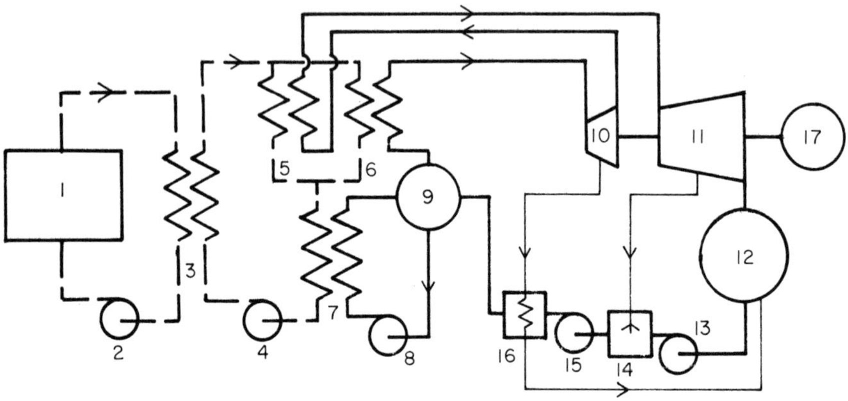

Fig. 4.9. Typical flow diagram for steam plant with recirculating boiler.

1. Reactor core
2. Primary sodium pump
3. Intermediate heat exchanger
4. Secondary sodium pump
5. Reheater
6. Superheater
7. Evaporator
8. Boiler circulating pump
9. Steam drum
10. High pressure turbine
11. Low pressure turbine
12. Condenser
13. Extraction pump
14. Low pressure direct-contact feed heater
15. Boiler feed pump
16. High pressure indirect feed heater
17. Alternator

Figure 4.9 shows a typical 2500 MW (heat) reactor with recirculating steam generators and Fig. 4.10 shows the heat transfer conditions. Steam is supplied to the turbine at 16 MPa and 500°C and reheated at 2.7 MPa to 500°C. The final feed temperature is 288°C, and the final stage of feed heating is done in the steam drum where the saturation temperature is 347°C. With a typical feed heating system the steam flow-rate is about 1070 kg s^{-1} through the superheaters and 750 kg s^{-1} through the reheaters. The total secondary sodium flow rate is 12000 kg s^{-1}. With a 4 to 1 recirculation ratio water enters the evaporators at a rate of 4280 kg s^{-1} and leaves with a dryness fraction of 0.35. The recirculation ratio is chosen to maintain nucleate boiling and therefore good heat transfer throughout the evaporators.

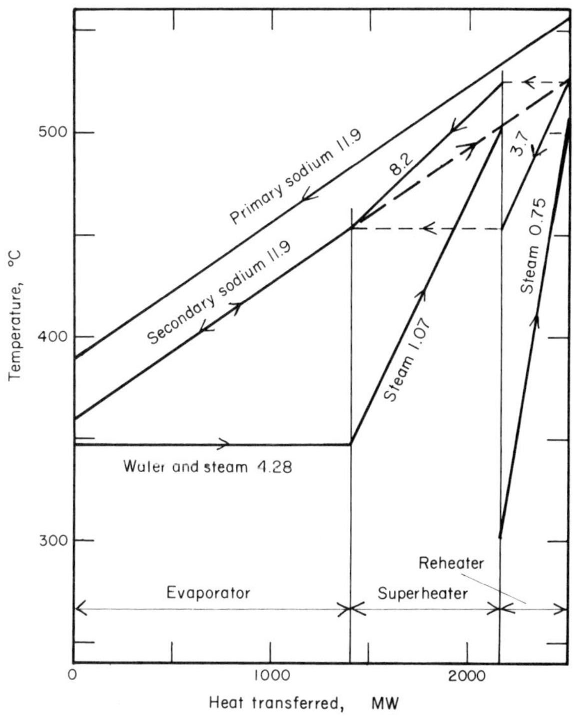

Fig. 4.10. Heat transfer in a 2500 MW power plant with steam at 500°C and 16 MPa at the turbine, and reheat to 500°C at 2.7 MPa. All flow-rates in Mg s^{-1}.

The choices available in designing the evaporators can be illustrated in a simplified way as follows. Suppose there are N evaporator tubes each of length L and diameter D. Then we have from equation 4.1

$$Q \simeq \overline{U}\Delta T_m N \pi DL, \qquad 4.2$$

where ΔT_m is the logarithmic mean temperature difference and Q is the total heat transfer rate in the evaporators (1400 MW in the reactor of Fig. 4.9). $\overline{U}$ is the mean heat transfer coefficient which depends on the steam-side conditions. Expressions for the heat transfer coefficient for nucleate boiling in forced convection are given by Collier (1972). The local value of $\overline{U}$ depends on D, the tube mass flow-rate m, and the local dryness x. The mean value over the range $0<x<0.35$ for $D = 0.02$ m is shown in Fig. 4.11 as a function of m.

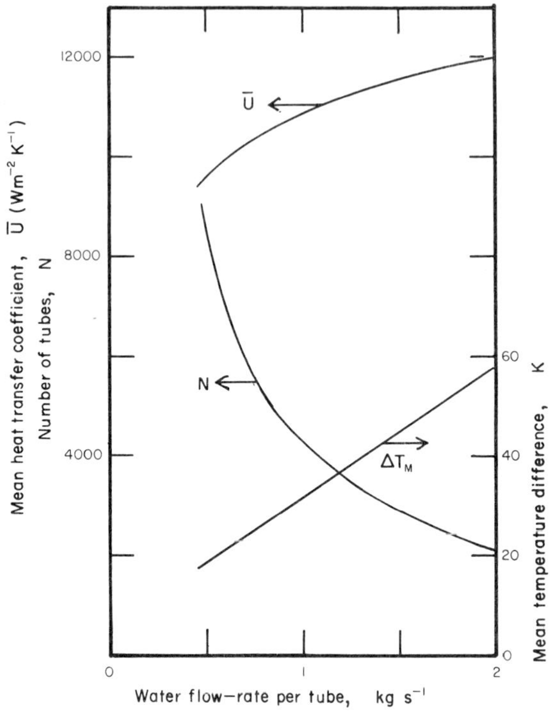

Fig. 4.11. Design options for a sodium-heated evaporator transferring 1400 MW to water boiling at 16 MPa with exit quality 0.35, total flow-rate 4280 kg s^{-1}, tube diameter 20 mm.

If the total mass flow-rate in the evaporators is M, then clearly $M = m\,N$, and equation 4.2 can be rearranged to give

$$L\Delta T_m = Qm/U\pi MD \qquad 4.3$$

The variation of N and ΔT_m with m for $L = 15$ m are also shown in Fig. 4.11.

D has to be at least 15 - 20 mm for ease of manufacture. For a 20 mm tube, $m = 1$ kg s^{-1} makes the mean speed of the two-phase mixture at the outlet about 12 m s^{-1}.

If m is much greater than this the problems of vibration and erosion are significant. This gives $N \simeq 4300$ and $\Delta T_m = 32$ K for $L = 15$ m. The main choices in design are thus between high m (vibration) or high N (large evaporators), and then between high L (long evaporators) or high ΔT_m (poor steam conditions). N = 4300 tubes could be accommodated in, for example, 6 separate evaporators, each containing 720 tubes in a bundle about 0.8 m in diameter.

The design choices for the superheaters and reheaters are very similar. U is smaller for the single-phase superheated steam, (typically 5000 W m^{-2} K^{-1} in the superheater and 2000 W m^{-2} K^{-1} in the reheater). Q is also lower, however, and, as an illustration, ΔT_m = 50 K can be achieved with 6400 tubes 8.7 m long in the superheaters and 4400 tubes of the same length in the reheaters give ΔT_m = 72 K. A typical arrangement might thus have 12 separate superheaters and 6 reheaters.

4.3.3 "Once-Through" Steam Generators

The main advantage of the recirculating steam generator shown in Fig. 4.9 is that the evaporator tube walls are always covered with water. The flow of the two-phase mixture of water and steam is either "bubbly" (near the water inlet and where the steam is in bubbles dispersed throughout the water) or "annular" (near the outlet where the steam bubbles coalesce to form a continuous vapour region in the centre of the tube and most of the water is in a film on the wall). The boiling water on the wall gives a high heat transfer coefficient so that the wall temperature stays close to the saturation temperature.

It also has the advantage that the mass of steam and water in the steam drums tends to decouple the reactor and sodium coolant circuits from rapid changes in demand for steam caused by fluctuations in the electrical load, so that control of the plant is easier. But it has the disadvantage of being complex and expensive.

A "once-through" steam generator shown in Fig. 4.12 is much simpler and cheaper because it does not require steam drum or boiler circulating pumps. The feed water enters a single heat exchanger in which it is heated to saturation, evaporated, and then superheated. The disadvantage is that somewhere along the tube the wall ceases to be covered with water. (This is either the point of "departure from nucleate boiling", "DNB", where nucleate boiling gives way to film boiling, or the "dryout" point, where annular flow gives way to dispersed flow (Collier, 1972).) At this point the heat transfer coefficient falls substantially.

If the tube is heated by a high temperature source such as a flame or a hot gas, as it would be in a fossil-fuelled boiler, the heat flux is approximately uniform, so the fall in heat transfer coefficient causes a sharp rise in the tube temperature and thermal stresses are generated. As the DNB or dryout point fluctuates up and down the tube is subject to transient thermal stresses which cause fatigue. But if the tube is heated by sodium which is probably not more than 100 K hotter than the steam the fall in heat transfer coefficient causes a drop in the local heat flux and only a modest rise in the tube temperature. Thus once-through steam generators may be more appropriate for sodium-cooled reactors than gas-cooled reactors or fossil-fuelled power plants.

It is very difficult to engineer sodium-heated reheaters with a once-through steam generator. This is a disadvantage as it reduces the thermal efficiency of the plant because the mean temperature at which heat is transferred to the steam is lower, and also as the wetness at the low-pressure end of the turbine is increased. There is little that can be done about the reduced efficiency, but the wetness can be reduced by means of moisture separators between the turbine stages, or by employing bled-steam reheat as shown in Fig. 4.12. This takes some of the steam from the high-pressure turbine and uses it to heat the main flow of steam after further expansion. The bled steam is condensed in the reheater but it is still hot enough to be used in a feed heater. There is a loss of thermal efficiency because of the entropy increase in the reheater, but this may be offset by the increased efficiency of the final turbine stages due to the lower wetness.

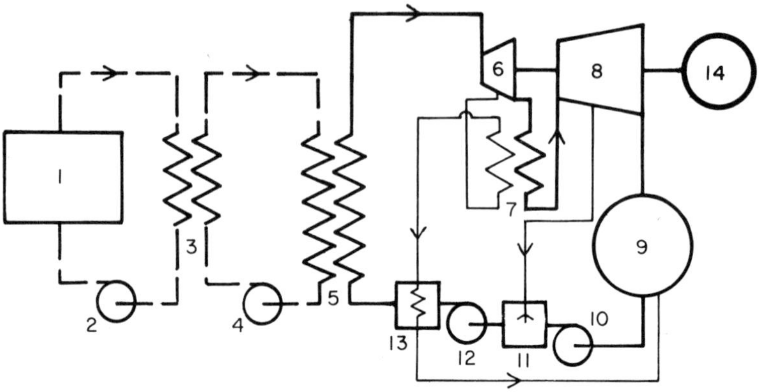

Fig. 4.12. Typical flow diagram for steam plant with once-through boiler.

1. Reactor core
2. Primary sodium pump
3. Intermediate heat exchanger
4. Secondary sodium pump
5. Steam generator
6. High pressure turbine
7. Reheater
8. Low pressure turbine
9. Condenser
10. Extraction pump
11. Low pressure direct-contact feed heater
12. Boiler feed pump
13. High pressure indirect feed heater
14. Alternator

4.3.4 Steam Generator Leaks

If the risk of a leak is to be accepted a steam generator has to be designed so that the consequences of a leak can be kept within bounds. The chemical reaction

$$2\ Na + 2\ H_2O \rightarrow 2\ NaOH + H_2$$

is exothermic and if it takes place at 350°C with the water as liquid the enthalpy of reaction is about 8 MJ per kg of water. If the reaction were adiabatic the products would reach a temperature of about 1850°C in steady flow.

Formation of sodium hydroxide is not the whole story, because some sodium oxide is formed by the reaction

$$2\ Na + H_2O \rightarrow Na_2O + H_2.$$

The extent to which each of these reactions takes place in the non-equilibrium conditions of a steam leak is not known, but it is clear that there is a significant potential for damage.

If there is a large leak, such as might be caused by a steam tube breaking in two, water or steam would be ejected into the sodium at a rate of the order of 1 kg s^{-1}, and the secondary sodium circuit would be exposed to at least the steam pressure of 16 MPa or so, possibly higher pressures as the sodium and water react. It is essential that the intermediate heat exchangers should be protected from damage because they form part of the barrier which contains the radioactive primary sodium.

This is normally done by fitting the secondary sodium side of the steam generators with bursting discs. If a large leak occurs the reacting sodium and water are expelled through the bursting discs into an effluent system which traps the products. At the same time fast-acting valves are closed to isolate the heat exchanger on both the steam and sodium sides. It may be that damage would be done to the affected unit but the rest of the plant would be protected.

If the leak is slightly smaller it may be possible to isolate the affected unit if the valves are triggered by a pressure sensor in the sodium side without invoking the bursting discs. This has the advantage that the task of removing reaction products from the effluent system is avoided.

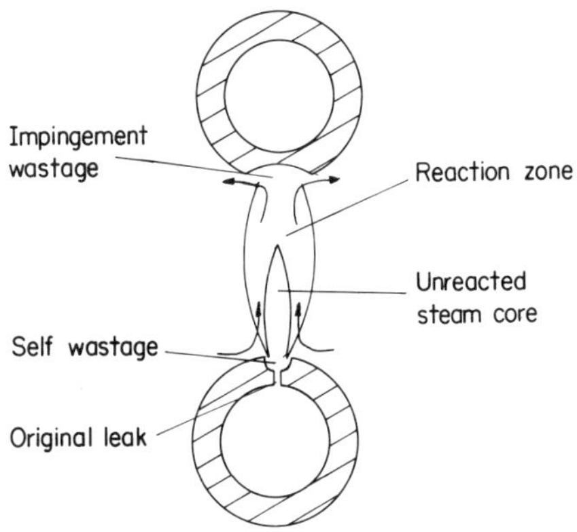

Fig. 4.13. Structure of a reacting steam jet in sodium and wastage of steam tubes.

A leak such as might come from a crack in a weld would be very much smaller and would not generate high pressures. If the crack is equivalent to a circular hole of diameter 0.1 - 0.2 mm the leak rate is 10 - 100 mg s^{-1}. The jet of steam or water reacting with the sodium forms a reaction zone rather similar to a flame (Tregonning, Mackay, and Buxton, 1975) with high temperatures at the tip. If this jet impinges on the adjacent tube it causes it to waste away by a mixed process of high temperature erosion and corrosion. At the same time the material round the original hole is wasted away in a similar manner. The processes are shown in Fig. 4.13. As a result the leak increases, both as the hole grows increasing the leak rate, and as further holes are formed in other tubes. The wastage rate of adjacent tubes as a function of leak rate is shown in Fig. 4.14. Wastage of various materials is discussed by Anderson (1979).

Very small leaks, of the order of 0.1-1 mg s^{-1} from holes of equivalent diameter 0.01 - 0.02 mm, have a tendency to become blocked instead of growing. After being blocked for a time such a leak can reappear suddenly and grow rapidly to 0.1 - 1 g s^{-1} (Smedley, 1976).

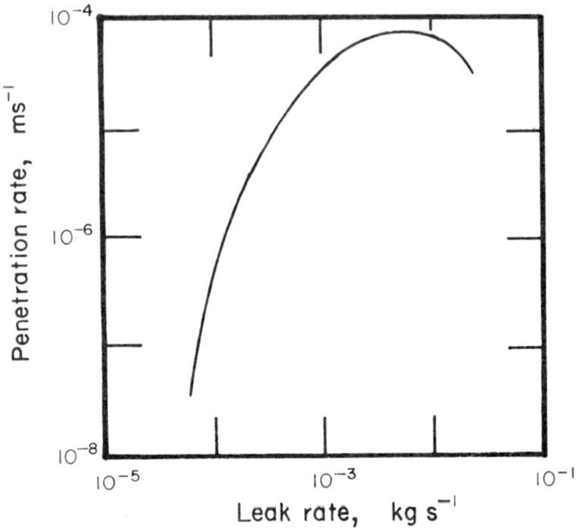

Fig. 4.14. Leak propagation: penetration rate through 2.25 Cr 1 Mo steel vs water leak rate (Smedley, 1976).

4.3.5 Leak Detection

Because leaks can increase it is necessary to detect them quickly and repair them before they cause widespread damage. The most widely-used method of detection is to observe the hydrogen by allowing it to diffuse through a nickel membrane as described in section 4.2.7 above. If leak detection by measuring an increase in the hydrogen concentration is to be sensitive, the hydrogen concentration has to be low to start with. This means that the build-up of hydrogen in the secondary sodium from oxidation of the water side of the heat exchanger tubes has to be controlled by cold-trapping.

Hydrogen leak detection has the disadvantage that it is difficult to design a system to respond quickly, in less than 10 s or so. Acoustic methods may provide a means of detecting leaks more quickly. Hans and Dumm (1977) survey in considerable detail the various methods of detecting leaks.

Once a leak is detected it has to be repaired. The affected unit has to be isolated and opened, and then the leak must be located. In a heat exchanger containing 1000 tubes this is no small task. Acoustic methods may again be useful, for if one side of the unit is pressurised with gas it may be possible to hear the gas issuing from the leak. An alternative method which has proved effective (Broomfield and Smedley, 1979) is to seek sodium hydroxide on the water side of the tubes by chemical means, for it is found that sodium migrates through small leaks against the pressure difference. If the leak is above the sodium level it is necessary to raise the liquid level to wash the region of the leak before the unit is opened and while it still contains steam.

When the leak has been located the usual method of repair is to plug the affected tube or tubes. The loss of heat transfer area is a small price to pay for quick return of the unit to operation.

In austenitic heat exchanger units there is a danger of more extensive damage because of the susceptibility to caustic stress corrosion cracking. If after manufacture part of the unit such as the tube plate is left in a state of stress, and there is a leak nearby, the sodium hydroxide formed may cause extensive cracking of the stressed region (Broomfield and Smedley, 1979). This is an important disadvantage of austenitic steels in steam generators, and is one of the reasons for the increasing popularity of ferritic materials.

4.3.6 Steam Cycle Design

The steam cycle in Fig. 4.9 is very similar to that of a conventional fossil-fuelled power plant, because the maximum steam temperatures are similar. The similarity is to some extent fortuitous, however, because the maximum temperature is set by different considerations in the two cases.

In a fossil-fuelled power plant the maximum steam temperature is limited to about 565°C because anything higher would require the use of austenitic steel rather than ferritic, and the increased cost would not be justified by the gain in efficiency. In a sodium-cooled fast reactor plant, however, austenitic steel is often widely used, as we have seen. The main temperature limits are the maximum permissible fuel cladding temperature, and the temperature of the structure above the core in a pool reactor or of the hot-leg pipework in a loop reactor. The parts of the structure in contact with hot primary sodium are subject to substantial thermal shock if the reactor is shut down suddenly in emergency, and the primary sodium maximum temperature is limited to a level at which this can be withstood.

The problems of withstanding thermal shocks has even led some designers to propose reduction of the primary sodium maximum temperature to below 500°C (Horst, 1978; Anderson, 1978). Steam superheaters are omitted, and saturated steam from the steam drum is fed to the turbine. Moisture separators and reheaters (using steam from the drum) are incorporated between some of the turbine stages to avoid the irreversibilities associated with high moisture content in the steam. The resulting cycle is very similar to that of a boiling water reactor.

If maximum steam temperatures of 500°C or above are permissible, a conventional steam cycle with superheat and reheat can be used, as shown in Fig. 4.9. The details of the cycle, and in particular of the feed heating system, may be slightly different from those of a fossil-fuelled power plant because the effect of final feed temperature on efficiency is rather different if heat is being transferred from a relatively low-temperature coolant than if it is transferred from high-temperature flue gas. The point is discussed in detail by Haywood (1975) and can be illustrated by reference to Fig. 4.15.

The cycle shown in Fig. 4.9, in which the final stage of feed heating is accomplished by mixing in the steam drum, is possible only because the saturation temperature (which is the lowest temperature on the steam side of the steam generators) is less than the minimum secondary sodium temperature. If it were sought to increase the cycle efficiency by increasing the steam pressure without changing the secondary sodium temperatures, it would be necessary to incorporate an economiser section in the steam generator (see Fig. 4.15 B). This would serve to cool the secondary sodium after leaving the evaporator, and would heat the feedwater from a new final feed temperature, which must be less than the minimum secondary sodium temperature, to the new, higher, saturation temperature. As a new feed temperature would be needed, the feed heating system would be different, showing the inter-dependence of feed and saturation temperatures in a plant of this type. In a fossil-fuelled plant, they are independent (Haywood, 1975).

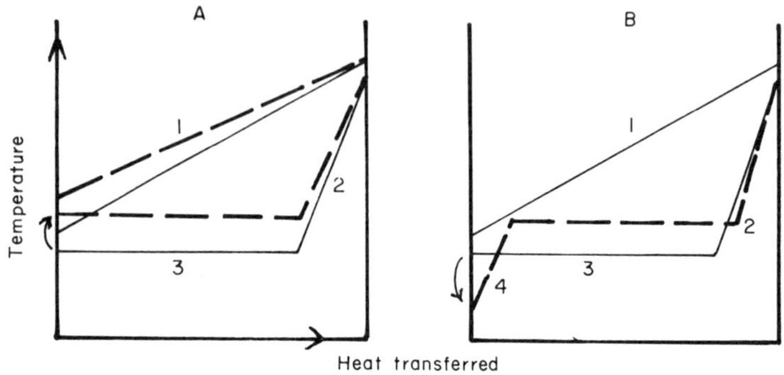

Fig. 4.15. Alternative ways of increasing efficiency by increasing steam pressure.

A. Increasing feed temperature by increasing primary and secondary sodium flow-rates.

B. Decreasing feed temperature by incorporating a sodium-heated economiser.

1. Secondary sodium
2. Superheater
3. Evaporator
4. Economiser

In this example, the increased saturation temperature could, of course, be accommodated by allowing the minimum secondary, and also primary, sodium temperatures to rise. This in turn would require an increase in the secondary and primary sodium flow rates (Fig. 4.15 A), and would involve the disadvantages of higher coolant speeds, more likelihood of vibration of heat exchanger tubes and fuel elements, greater pressure differences and stresses in the core, bigger circulating pumps and so on.

4.3.7 Plant Efficiency

The use of the secondary sodium circuit to separate steam from radioactive primary sodium carries the disadvantage of a loss of thermal efficiency due to the increase of entropy as heat is transferred from primary to secondary sodium. The extent of the loss can be estimated very conveniently in terms of the flow of exergy, or "steady flow availability" as it is sometimes called. The theory is explained by Haywood (1975).

If the working fluid in a cyclic power plant circulates at a rate $\dot{M}$, and receives heat as it changes from state 1 to state 2, the rate at which it receives heat is Q, where

$$Q = \dot{M}\ (h_2 - h_1), \qquad 4.4$$

and the maximum rate of work output, P_{max}, is given by

$$P_{max} = \dot{M}\ (\epsilon_2 - \epsilon_1). \qquad 4.5$$

Here ϵ is the specific steady flow exergy, defined by $\epsilon \equiv h - T_0 s$ where h and s are specific enthalpy and entropy respectively, and T_0 is the dead-state or environmental temperature. The maximum thermal efficiency is P_{max}/Q. P_{max} is attained only if all the processes taking place in the plant are reversible. Equation 4.5 shows that P_{max} is the rate of increase of exergy of the working fluid as it receives heat.

In the reactor illustrated in Fig. 4.9 Q is 2500 MW. The rate of increase in exergy of the primary coolant as it flows through the core is 1490 MW, while that of the secondary coolant in the intermediate heat exchanger is 1445 MW. The difference, 45 MW, or 1.8% of the heat transferred in the core, is the work output lost as a result of the irreversibility of the heat transfer.

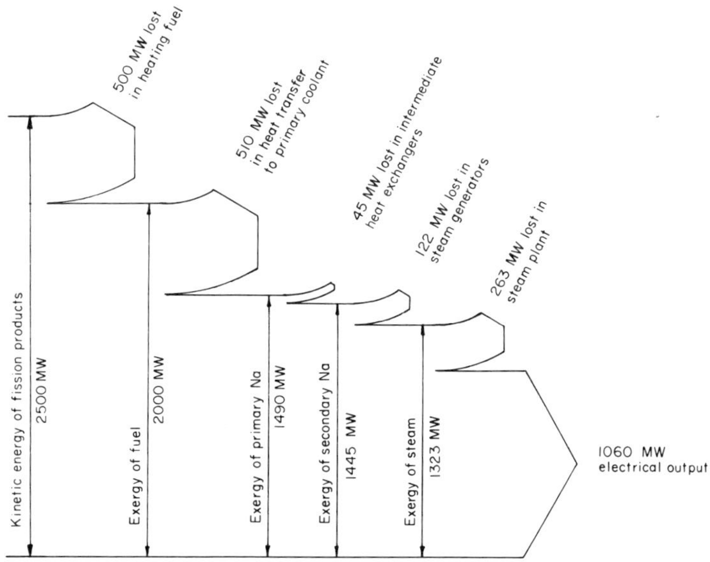

Fig. 4.16. Exergy flow in the reactor and power plant of Figs. 9 and 10. The losses are lost opportunity for work output due to thermodynamic irreversibilities.

The increase in exergy of the steam in the steam generator is 1323 MW, and this would be the nett power output if the steam cycle were reversible. There are, however, various irreversibilities, in the turbine for example, due to pressure losses in pipes, and in the feed train (the irreversible mixing in the steam drum causes a loss of 9 MW) and the actual nett work output would be about 1060 MW, giving an overall thermal efficiency of 42%. Thus it can be seen that the loss of work output due to the introduction of the secondary coolant circuit is relatively unimportant. The flow of exergy in the plant is shown diagrammatically in Fig. 4.16.

4.4 CONTROL SYSTEMS

4.4.1 Normal Operation

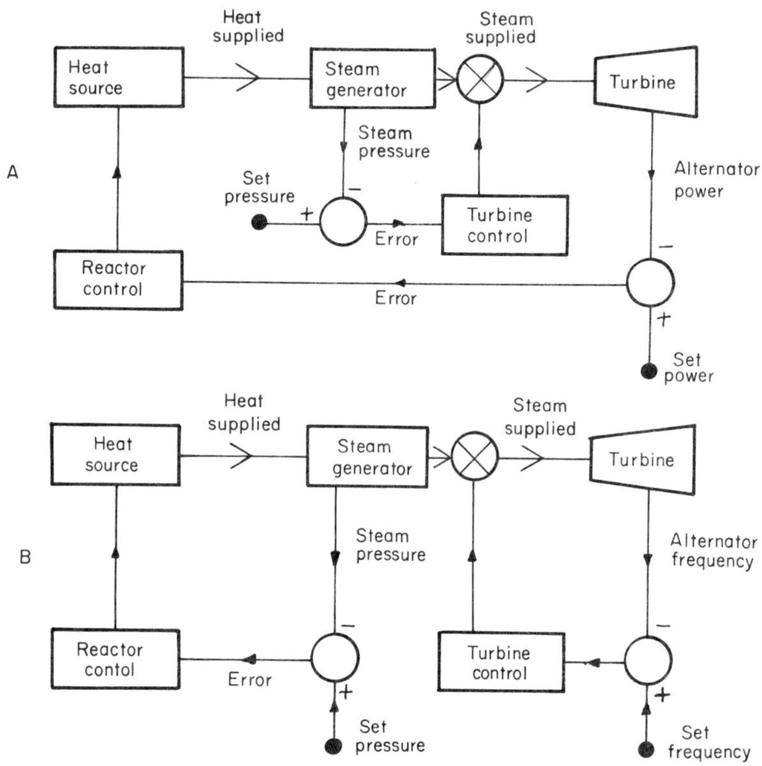

Fig. 4.17. Plant control systems.

A. Base load
B. Load following

There are two main strategies for controlling a power plant of any sort, which are chosen according to whether it is required to generate constant power or to respond to a varying demand. The former is adopted for base-load power plant, and because of their low marginal fuel cost nuclear power stations are often used in this way. If the nuclear generating capacity exceeds the minimum demand from the system however, some nuclear power stations have to be operated in the latter "load-following" manner.

The appropriate control systems for base load and load-following operation are shown in outline in Fig. 4.17. The "heat source" in the case of a fast reactor power station consists of the reactor together with the primary and secondary sodium circuits. If base-load operation is required, deviations of the power generated by the alternator from the preset required value are used to control the reactor heat output. The turbine control valve is then adjusted to keep the steam pressure at a preset level. If for some reason the reactor power should rise, it causes an increase in the steam pressure. The turbine control valve opens to reduce it and so increases the power generated by the alternator, and this in turn causes the reactor power to be reduced.

If load-following is required the turbine control valve is adjusted to keep the alternator frequency equal to the required preset level. Deviation of the steam pressure from a preset level is then used to control the reactor power. If the load demanded increases the alternator is very slightly and briefly retarded. The reduction in alternator frequency is used to open the control valve to admit more steam and return the alternator to its correct frequency and phase. This reduces the steam pressure, and the control system acts to increase the reactor power and restore the pressure to its preset value. The principles of these control systems are described by Knowles (1976).

Figure 4.18 shows the control of the "heat source" in a little more detail. One way to adjust the heat output to give the required power output (for base load) or steam pressure (for load-following) is to use the control variable to control the primary and secondary pump speeds. The reactor control rods are then moved to keep the core outlet temperature at a preset level. If more power is required the primary and secondary pump speeds are increased in step. The increased flow through the core decreases the outlet temperature, and the control rods are adjusted to increase the reactor power. A typical fast reactor control system is described by Evans and others (1967).

The various control system elements shown as boxes in Figs. 4.17 and 4.18 are not just simple proportional controllers. Derivative or integral terms have to be included in the control function to make the system stable and the overall gain has to be chosen to keep the main operating parameters within acceptably narrow limits. This is particularly true of the coolant temperatures, because fluctuations of the sodium temperature are transmitted so readily to the structure (see section 3.3.7).

In normal operation it is not very difficult to keep temperatures constant. Because of the large mass of sodium in the primary circuit, especially in a pool reactor, power or coolant flow rate fluctuations cause only very slow temperature changes in most of the structure. The exception may be any structural members, such as control rod guides or mechanisms, exposed to the coolant immediately on leaving the core (section 3.3.7). Similarly the mass of water in the steam generators and particularly in the steam drums of a recirculating steam generator, and the mass of metal in the drums and heat exchangers, tend to smooth out variations in steam demand so that the steam pressure responds quite slowly to changes in the turbine control valve opening.

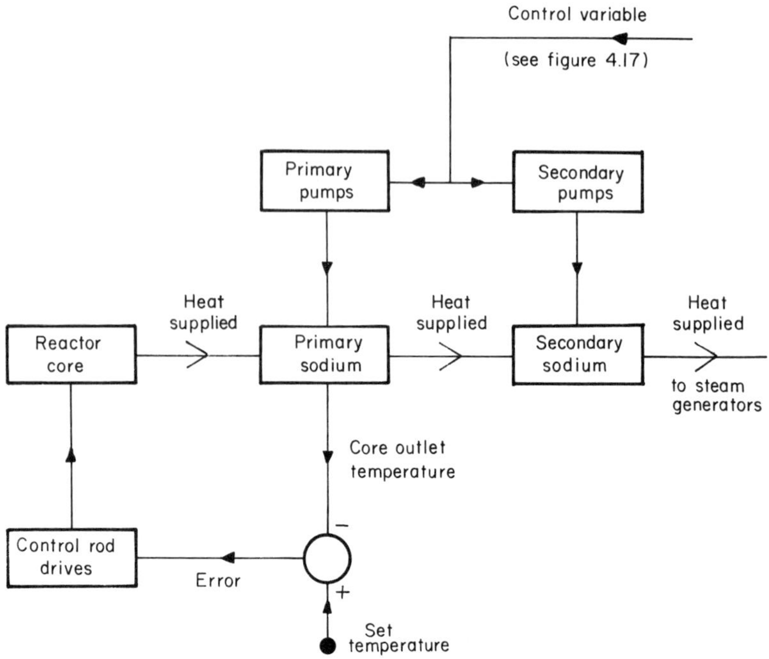

Fig. 4.18. Reactor heat source control system

There is an important difference, however, between the steam plant of a sodium-cooled fast reactor and that of a fossil-fuelled power station. Because of the much lower heat transfer coefficients in a fossil-fuelled station where the heat is transferred to the boiler tubes by convection and radiation from a gas the boilers are very much larger, by a factor of 100 or more, than the sodium-heated steam generators of a fast reactor power station. This is not important in normal operation, but under "upset" conditions, especially when the steam plant is operating abnormally, the rapid response of the steam generators to pressure changes can create difficulties in control.

4.4.2 Abnormal conditions

There is a range of abnormal events which befall power stations so frequently that they cannot be though of as accidents even though they interrupt ordinary operation. It goes without saying that such events cannot be allowed to hazard the operating staff or the public. In addition the control system has to be designed so that they do not cause damage which would be expensive to repair, and so that the loss of electrical output which has to be made good from another more expensive source is kept to a minimum.

Examples of these abnormal events are failures of the connection to the power distribution system, which might be due to storm damage to a power line, and would cause a sudden loss of the alternator load. Failure of pieces of auxiliary equipment such as the bearing lubrication system or the alternator cooling system would demand rapid shutdown of the turbine. Failure of the feed pumps, the boiler circulating pumps or the primary or secondary sodium pumps would prevent full-power operation of the plant and probably require complete shutdown. Perhaps the most important sources of such events are failures of the reactor and plant protective system itself. This is designed to be "fail-safe", so that for example if an ionisation chamber measuring neutron flux goes wrong it gives an indication of high flux. If two or more fail the protective system will take action as if the flux were really too high and shut the reactor down automatically. This is usually know as "tripping" the reactor.

The control system has to be designed so that the plant or the affected part of it can be shut down safely and without damage if one of these abnormal, or "upset", events happens. The design of the turbine and alternator trip system can follow conventional lines, but tripping the reactor and the sodium coolant circuits involves some novel considerations, the two most important of which are the removal of the decay heat and the avoidance of thermal shock.

Even if the reactor is shut down the fuel continues to generate heat by the decay of radioactive fission products. Emergency cooling systems are provided to remove this heat under accident conditions (section 4.2.2), but it may be preferable not to rely on them in an upset event. It may be better to maintain operation of the primary and secondary sodium circuits in case of a reactor trip, and to reject the decay heat to the condenser by maintaining the supply of feedwater to the evaporators. In the event of a turbine trip it may be best to open a bypass so that steam can flow directly to the condenser. If one primary or secondary sodium pump fails it may be decided to trip the reactor but keep the other pumps operating to remove the decay heat.

It is more difficult to avoid thermal shock. If the turbine trips either the full flow of steam has to be diverted to a very large standby "dump" condenser (which is expensive), or the reactor has to be tripped or reduced in power very quickly. If the latter course is taken the sodium flow-rates have to be matched to the power if rapid changes in temperature of structure above the reactor core and in the intermediate heat exchangers are to be avoided.

Considerations such as these impose considerable demands on the control system, especially as the time-constants for temperature changes in the reactor core and the steam generators are quite short.

CHAPTER 5

Safety

5.1 INTRODUCTION

The designer of a fast reactor, just like the designer of any other engineering enterprise, has to take into account what might happen if something goes wrong. He has to make sure that whatever happens the risk of injury, either to the operating staff or the general public, or of damage to property, is very slight.

There are basically two ways of making a reactor safe. First the overall design concept is chosen so that it is inherently safe. That is to say that for a number of possible accidents the design is such that the reactor behaves safely and damage does not spread even if no protective action, automatic or deliberate, is taken.

It is not possible to guard against all accidents in this way however well the overall design is chosen. The second way to make the reactor safe is to incorporate protective systems. These are devices designed specifically to prevent the damaging consequences of accidents. A protective system can be active, like an automatic shutdown system, or passive, like a containment barrier.

The overriding aim is to make sure that the risk to the public is extremely small, and to ascertain that the aim has been met the designer has to determine the response of the reactor with its protective systems to a range of accidents. To test the systems thoroughly it is often necessary to assume certain accidents happen even though no way is know by which they could actually take place. For this reason they are known as "hypothetical accidents".

The final step is to analyse all the accidents, whether hypothetical or not, and to ensure that the risks meet the criteria imposed by the authorities which regulate nuclear establishments. These criteria vary of course from country to country.

This chapter starts by describing the inherent features of a sodium-cooled oxide-fuelled fast reactor which make for safety, and goes on to discuss some of the protective systems which can be used. The development of a few of the most important hypothetical accidents are outlined briefly, and finally risks and their assessment are discussed. As before we concentrate on the aspects of reactor safety which are peculiar to sodium-cooled fast reactors. The safety of nuclear reactors in general is discussed by Lewis (1977) and Farmer (1977). Detailed accounts of some of the subjects touched on briefly here are given by Graham (1971).

The main concern in reactor safety is to make sure that the radioactive materials - fuel, fission products and activation products - are contained adequately and do not escape to the environment. There are other risks, associated with steam and electrical plant, for example, but as they are the same as for any other power station they are not dealt with here.

This chapter is confined to consideration of the safety of the reactor alone. Questions of fuel manufacture and transport or of waste disposal are not entered upon.

5.2 INHERENT SAFETY FEATURES

5.2.1 Containment Boundaries

There are several features of a sodium-cooled oxide-fuelled fast reactor which are advantages from the point of view of safety, and a few which are disadvantages. Chief among the advantages is the fact that the radioactive materials are separated from the environment by several substantial physical boundaries, which would have to be broken before they could be released.

There are usually three boundaries: the cladding of the fuel elements, the primary coolant containment (i.e. the reactor vessel in a pool layout, or the reactor, pump and heat exchanger vessels and the connecting pipework in a loop layout), and the reactor building. Fuel and fission products are contained by all three, activation products including ^{24}Na by the last two.

Failure of the fuel cladding is discussed in section 2.4.7. The fuel is designed so that failure of any one element is very unlikely, but there are so many elements in a reactor core (of the order of 10^5 in a 2500 MW (heat) reactor, each being renewed after each year or so of operation at full power) that the possibility of some failures has to be allowed for. Failure by fission-product corrosion or due to a defect in manufacture is likely to result in no more than a small crack in the cladding, which would release to the coolant just the fission-product gases and possibly some volatile fission products. The amount of radioactivity reaching the coolant would be small unless a much larger breach was made in the first place, or the small breach was enlarged by operating the reactor for a long time without replacing the failed fuel. Experiments have shown that even in the event of gross cladding failure the amount of fuel released is very small (Smith and others, 1978).

If there should be widespread cladding failure gaseous fission products would find their way to the cover gas over the coolant. They would be contained by the roof of the reactor vessel, but there might be some leakage through the seals on pump shafts, control rod actuators, rotating shields, etc. The seals have to be designed to keep this, together with the leakage of ^{24}Na, to a low level.

Apart from such minor leaks, the primary coolant containment also has to be designed to prevent a major breach. This means that it has to be able to withstand the effects of accidental loads which might be imposed from without, by a piece of heavy equipment being dropped, for example, or from within by a core accident (see section 5.4.4).

Any release of radioactive material from the primary coolant circuit is contained by the reactor building, or "secondary containment" as it is often known. The building has to have a ventilation system with suitable traps and filters to control any radioactivity released to the atmosphere inside the building and to cope with the effects of the sodium fire which would result from a breach of the primary or secondary coolant circuits. (Ways of preventing such a fire are described in section 5.2.4 below.) The building itself has to withstand loads imposed by the wind, by earthquakes (if the reactor is to be built in a zone where these occur), or by missiles from external sources such as an explosion in a nearby piece of plant or equipment.

In summary although care has to be taken in design it is reasonably straightforward to ensure that the three containment boundaries are independent and will remain intact under the loads imposed by a wide range of less severe accidents, which are the ones most likely to occur. This is the principal foundation of the safety of a reactor of this type. Once it has been established, release of radioactivity is possible only if there are coincident independent failures of all three boundaries, which is very unlikely, or if some single initiating event can cause breaches in all three.

5.2.2 Coolant Pressure

The second major advantage of a sodium-cooled reactor from the point of view of safety is that the coolant pressure is low, so that the primary coolant containment is only modestly stressed and is unlikely to fail, and if it should fail the coolant does not vaporise. It is possible to design the system so that in the event of a primary circuit rupture the core can be cooled without the provision of emergency supplies of coolant. This can be done by surrounding the vessels and pipework by a leak jacket, and by placing the reactor vessel in a hole in the ground.

This protects a pool reactor against failure of both vessel and leak jacket because the primary coolant cannot fall below the level of the core and the intermediate heat exchangers. Decay heat can be removed indefinitely by the secondary coolant provided the intermediate heat exchangers are intact or by means of an emergency heat rejection system. In a loop system the coolant pipes have to be connected to the reactor vessel above the level of the core if the core is to remain covered in the event of a pipe break. Decay heat can be rejected if at least one of the primary coolant circuits remains intact.

It is normal to provide auxiliary pump motors to guarantee circulation of the primary coolant at a sufficient rate to remove the decay heat even if all the main pump motors or the electrical supply to them should fail. This may not actually be necessary, however, because natural circulation may be quite adequate to remove decay heat from the core without excessive overheating.

Thus there are two advantages of the low coolant pressure: that the coolant containment is unlikely to be broken, and, even if it is, it is still quite easy to remove the fission product decay heat and keep the cladding cool enough to prevent widespread failure. This is in complete contrast to gas and water-cooled reactors where the coolant pressure is high and extensive protection has to be provided against loss-of-coolant accidents.

An additional safety feature, particularly in pool reactors, is provided by the large mass of the primary coolant, which may exceed 2000 tonnes. Provided it circulates, by means of pumps or natural convection, its temperature rises only slowly even if there is no secondary cooling at all. This is illustrated by Fig. 5.1, which shows the rise in the mean primary coolant temperature assuming the reactor shuts down and simultaneously all secondary cooling is lost. There is clearly plenty of time to make secondary cooling available before the temperature rises enough to cause widespread fuel failure (which might happen in the range

800 - 1000°C, depending on the details of the fuel design and the burnup). At atmospheric pressure sodium boils at 892°C.

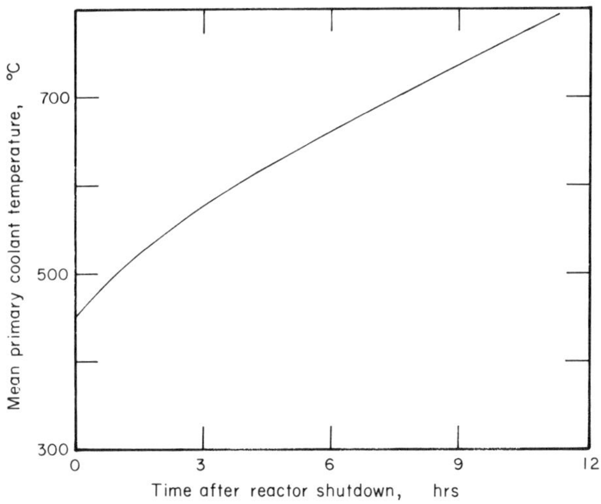

Fig. 5.1. Mean primary coolant temperature in a typical loop reactor after reactor trip and coincident total failure of the heat rejection system.

5.2.3 The Negative Power Coefficient of Reactivity

The third important safety advantage is that the power coefficient, of which the Doppler effect is an important component, gives a reliable prompt negative feedback of fuel temperature on reactivity.

Reactivity is normally controlled by moving the control rods, and provision has to be made to prevent damage if the control rod mechanisms should go wrong and withdraw the rods inadvertently. The unplanned rise of power can be detected by several independent means such as neutron flux or coolant temperature, and the reactor can then be shut down by inserting the safety rods, which operate independently of the control rods. The control rod mechanisms are usually made so that it is impossible to withdraw the rods at more than a slow speed under any circumstances.

In the very unlikely event that all the protective mechanisms should fail the power does not rise exponentially. The negative power coefficient of reactivity (see section 1.6.6) ensures that it rises slowly at a rate determined by the movement of the control rods. If in spite of all precautions the power continues to rise, however, the fuel might begin to fail. Under these circumstances, if fuel elements lose their integrity, it may no longer be possible to argue that the part of the power coefficient which depends on expansion of the cladding is reliable. The importance of the Doppler effect is that whatever happens to the fuel it is a reliable source of negative feedback.

The precise part played by the Doppler effect depends on what accident is assumed to take place. In the most extreme case, if the reactivity were increased by some hypothetical mechanism so fast that heat transfer from fuel cladding had no time to take place, and prompt criticality were to be approached, the prompt feedback from the Doppler effect prevents the power rising on a microsecond timescale determined by the prompt neutron lifetime. This point is discussed further in section 5.4.4 below.

5.2.4 Disadvantages

As well as these features making for the safety of the reactor there are some which do the reverse. The most important are the high power density, the possibility of an accidental reactivity increase, and the flammability of the coolant.

The high power density means that if the fuel is not cooled its temperature rises very quickly. In the extreme case of a hypothetical accident in which a central fuel element were somehow to be deprived completely of its cooling while the reactor was operating its temperature would rise at some 600 K s^{-1} and it would melt in 3 or 4 seconds.

If the fuel melts the possibility of rapid vaporisation of the coolant arises. It has been observed that if some molten metals, notably aluminium or steel, are mixed with cold water (as has happened in accidents at foundries) the resulting vaporisation is sometimes violent and a so-called "steam explosion" takes place. The mechanism is not properly understood, and the analogous "sodium vapour explosion" caused by mixing molten oxide fuel with liquid coolant has never been observed to take place with any significant severity. It may indeed be impossible for damage to be done by this mechanism, but until it has been proved the possibility has to be taken into account (Board, Hall, and Hall, 1975; Henry and Fauske, 1975).

The possibility of an accidental increase in reactivity arises (apart from failures of the control system mentioned in section 5.2.3) because the fuel is not in its most reactive configuration. As pointed out in section 1.6.2, if the fuel is made more compact the reactivity in general increases. This is however more of a theoretical disadvantage than a practical one because it is difficult to envisage a realistic means by which an accident could compact the core in the right way. The main reason is that there is not much reactivity to be gained if the core collapses vertically, especially for a core of large diameter (see Fig. 1.16). The reactivity would rise more if it were compacted radially, but it is very hard to postulate how this could happen.

The flammability of the coolant is included in the list of disadvantages for completeness, but it cannot be said to pose any significant danger. It is a feature which has to be taken account of in design, but this is relatively simply done by well-tried methods such as the provision of leak jackets and fire-fighting systems. In the most sensitive areas, such as the primary coolant circuit, it is usual to enclose all the sodium-carrying vessels and pipes in a continuous leak jacket. The space inside the leak jacket is filled with inert gas and provided with sodium detectors. In less sensitive areas such as the secondary coolant circuits leak jackets may be considered unnecessary, and safety ensured by provision of devices to catch any sodium which leaks and of equipment to deliver suitable extinguishing material. Mixtures of inert powders such as sodium chloride or sodium carbonate are often used.

The disadvantages of sodium are very small compared with the major advantages it possesses over water or gas (section 5.2.2).

5.3 PROTECTIVE SYSTEMS

5.3.1 Automatic Shutdown

The features described in section 5.2 are present whatever the details of the design. Protection is also given by systems designed deliberately to prevent accidents or prevent them from causing damage. Very often protective systems serve the dual function of preventing injury to people by stopping the release of radioactivity and of minimising damage to the reactor itself.

"Active" protective systems depend on detecting that something is wrong, and then taking automatic protective action which is usually to shut down the reactor. The output from an instrument, such as a thermcouple or a neutron monitor, is amplified and compared with a reference, or "trip", level. If the trip level is exceeded the protective action is taken.

If the trip system is to offer real protection a "fail-safe" system must be employed. This means that if the instrument itself fails the dangerous condition should be indicated. Thus if a thermocouple circuit is broken, or if it is short-circuited, the amplifier must give an output above the trip level. If the high voltage supply to a neutron detector fails so that it gives zero output the trip circuit must be activated.

It is of course essential to avoid tripping the reactor unnecessarily, and it must certainly not be tripped every time a thermocouple fails. This implies that a "two-out-of-three" system or a variant of it has to be used. Each of the instruments needed for protective action is triplicated. If one indicates danger an alarm is sounded but no other action is taken. If two or more indicate danger the reactor is tripped. This reduces the frequency of spurious trips due to instrument failures because they happen only if two fail at the same time. It also allows maintainence of the instruments while the reactor is operating. A detailed statistical discussion of the reliability of multiple protective systems is given by Lewis (1977), pp. 103 - 126.

The action taken automatically when the reactor is tripped is to insert the neutron absorbing rods into the core. It is essential that this insertion should be as reliable as possible. A typical arrangement is for the neutron absorber to be connected to the actuator, by which it is moved in normal operation, by an electromagnet. When the reactor is tripped the current to the magnet is interrupted and the rod falls into the core under gravity.

5.3.2 Whole-Core Instrumentation

The instrumentation needed to detect an incipient accident can be divided into two classes: that for accidents which affect the whole of the core (such as reactivity changes or primary coolant pump failure); and that for accidents which affect initially only part of the core (such as a coolant flow failure in one subassembly). The latter is discussed in sections 5.3.3 and 5.3.4 below.

It is relatively easy to detect an accident affecting the whole core by means of the instruments used to control the reactor in normal operation. The condition to be guarded against is overheating, so it is necessary to monitor the reactor power and the coolant temperature and flow rate.

The instruments which measure neutron flux (section 3.5.3) are used to provide a high power trip. A minor disadvantage of this is that the relationship between the reactor power and the flux at a measuring instrument in the reactor shield changes with time as the control rods are withdrawn to compensate for burnup and as plutonium builds up in the breeder, and depends on the pattern in which new subassemblies are loaded when the reactor is refuelled. Thus a high power trip based on neutron flux cannot be very precise, but it is very reliable.

The signal from the neutron flux instrumentation is also used to determine the inverse period $\partial C/C\partial t$, where C is the indicated flux level (see section 1.2.5). The inverse period is closely (but indirectly) related to the net reactivity, and the reactor is tripped when it becomes too large. This trip system is of little importance when the reactor is operating at power because feedback keeps the net reactivity close to zero. But when the power is very low, so that very little heat is generated, there is no feedback. Under these conditions an inverse period trip is a protection against accidental increases in reactivity.

Coolant flow can be monitored by electromagnetic flowmeters at the core outlet or by observing the rotation of the circulating pumps. It is normal to trip the reactor if the pumps stop or if the flow out of the core falls below a set value. If the reactor is to be operated efficiently at less than its full power, however, either the trip level on coolant flow has to be set low (at say 10% of the full flow rate) or the trip level has to be altered according to the power required.

This disadvantage can be avoided if the coolant temperature at the core outlet is used to trip the reactor. This is subject to some variation due to burnup and control rod movement during steady operation, but has the advantage of depending directly on the quantity (the temperature) which must not be allowed to become too high.

An alternative, which is useful as a second independent protection, is to calculate the ratio of measured power to measured flow rate and use this as a trip signal.

5.3.3 Subassembly Instrumentation: Failed Fuel

It is more difficult to detect an incipient accident affecting only one subassembly than one affecting the whole core, but it is necessary to do so because damage can be prevented from spreading if the reactor is shut down soon enough (see section 5.4.2). This means that the consequences of the accident and the cost of repair are much smaller. The difficulty arises because there are hundreds of subassemblies in the core and breeder. If instruments have to be attached to each, and triplicated, the resulting trip system is very complex. It may also be very cumbersome, especially if instruments have to be fixed to each subassembly after it has been inserted into the core and detached before it is removed.

Thus instruments outside the core which can detect a developing accident anywhere in the core are very desirable. There are two main candidates: failed cladding detectors and boiling detectors.

A failed cladding detection system searches for failed fuel elements by monitoring the coolant or cover gas for radioactive fission products which must have come recently from the fuel. This can be done in two ways. β or γ activity can be sought if the fission products are separated from the coolant, which is already γ-active. Alternatively neutrons from delayed neutron precursors can be sought if a sample of the coolant is removed from the neutron flux.

A typical system has four parts. A sample of the main coolant flow from the core is piped away to a point at which the neutron flux is low. There it passes to a vessel surrounded by moderating material and thermal neutron detectors (usually boron trifluoride counters) which detect any delayed neutron precursors in the coolant. This serves to detect clad failures anywhere in the core, provided the sample is representative of the flow through the whole core. This means that it has to be taken from a point where the flow from the core outlet has become well mixed.

Simultaneously samples are taken from the outlet of each subassembly in turn to a delayed neutron monitor. This serves to locate the subassembly in which a failure has occurred. Because there are so many subassemblies to sample location is relatively slow. A quicker response is given by the bulk sampling system.

Thirdly a sample of the cover gas over the sodium in the reactor vessel is taken to a γ-spectrometer and a moving-wire β-precipitator. This uses a charged wire to precipitate the daughter products of β-decaying fission products. The wire then moves to a chamber containing a β-detector. The system thus discriminates against a range of activation products which might be present, and selects β-active daughters of gaseous β-active fission products (mainly ^{88}Rb produced by β-decay of ^{88}Kr), so that it responds selectively to cladding failures. Further information about the nuclides present is given by the γ-spectrometer.

Finally the gaseous fission products are stripped from the location system coolant sample by a stream of argon which then goes to a β-precipitator. This provides an independent way of locating failed cladding.

5.3.4 Subassembly Instrumentation: Coolant Boiling

Failed fuel detection has the advantages that it is reliable and a direct indication of the release of radioactivity which is what basically has to be avoided. Even a bulk detection system is however relatively slow, taking some tens of seconds to detect a failed fuel element. This is adequate to control the release of radioactivity to the coolant, but a faster system could help to minimise damage to the fuel. The delay is due mainly to the time taken to transport the coolant sample to the detector.

If fuel element failure is caused by overheating it may be accompanied by boiling of the coolant (which takes place at about 920 - 940°C at the pressure in the core). An acoustic boiling detector might give a quicker indication that something is amiss than waiting for the fuel to fail and then for the failed fuel detection system to operate. The use of sound to detect boiling is suggested by the ease with which boiling is heard in a domestic kettle. It would be particularly attractive in a reactor because a small number of detectors might be enough to detect boiling anywhere in the core.

The main difficulty is that the reactor itself is quite noisy. Sound is generated by the coolant pumps, by the turbulence of the coolant flow, and by cavitation. Cavitation is particularly awkward because it is a form of boiling (caused by local reductions in pressure at points where the flow of the coolant is accelerated, such as at sharp corners or on the blades of the pump impellors), and it makes a very similar noise to boiling caused by overheating. It may be possible to avoid the difficulty by designing the reactor to keep cavitation to a minimum and to discriminate against other background noises by listening in a frequency range in which boiling generates a lot of noise.

In most cases boiling in a subassembly can be detected by means of a thermocouple at the outlet, but there may be some circumstances in which detection would not be reliable. A partial blockage to the flow somewhere in the subassembly could be large enough to cause severe overheating in its wake, possibly to the boiling point, but at the same time have a very small effect on the total flow rate and the mean outlet temperature. This is because the resistance of the subassembly to coolant flow is already high and the additional resistance caused by a blockage is small in comparison and reduces the flow rate only slightly. The effects of blockages are discussed in more detail in section 5.4.2 below.

It may be possible, however, to detect a local blockage by observing temperatures fluctuations at the subassembly outlet. A blockage increases the turbulence of the flow and the differences in temperature between different parts of the flow, and so causes increased temperature fluctuations, or "temperature noise", at the outlet. This can be detected with fast-response thermocouples, but it has not so far been used as a basis for a reactor trip system.

5.3.5 Post-Accident Heat Removal

So far we have discussed protective systems aimed at detecting an accident in its early stages and preventing it from developing. Another type of protective system is that which mitigates or controls the consequences of an accident. The reactor containment building with its ventilation and filtration system (section 5.2.1) can be regarded as one such. Another is the post-accident heat removal system.

The rate at which "decay heat" is generated by the decay of radioactive fission products in the fuel is shown in Table 5.1. Depending on the design it is probably possible to lose a few megawatts from, for example, the secondary coolant pipework by natural convection to the atmosphere. But reliable provision has to be made to remove heat for a few weeks until this level is reached, and for longer if the accident has made the secondary coolant circuits inoperable.

TABLE 5.1 Decay Heat Produced by Radioactive Decay of Fission Products after Shutdown from Steady Reactor Operation for an Infinite Period

Time after Shutdown	Fraction of Reactor Power before Shutdown
1 second	.062
10 seconds	.050
100 seconds	.035
1 hour	.015
1 day	.0045
1 week	.0019
1 month	.0011
1 year	.00056
10 years	.00026

This is done by providing independent emergency heat rejection circuits with separate heat exchangers in the reactor vessel. The coolant in these circuits circulates by natural convection to air-cooled heat exchangers. As sodium freezes at 98°C the coolant is usually a mixture of sodium and potassium which freezes below normal atmospheric temperature.

With such a heat rejection system the fuel can be cooled safely even if the secondary circuits do not work and there is no electrical power to drive the primary pumps. Natural convection within the reactor vessel transfers the heat to the emergency heat exchangers, and from them it is rejected to the atmosphere.

It is usually held that after an accident the coolant would be able to circulate through the core. It is sometimes argued however that it is necessary to guard against an accident which somehow destroys the core and breaks up the fuel elements into small pieces which fall in a heap at the bottom of the vessel, or melts the fuel so that it falls to the bottom. The decay heat generated in the heap might be sufficient to melt the bottom of the vessel and so violate the containment. It is also conceivable that the heap might attain the right configuration to form a critical mass. Experiments to determine the behaviour of such a system have been reported by Plein and others (1979).

If it is accepted that these concerns are realistic, a device can be provided which will catch fragments of fuel and support them in a configuration which will allow them to be cooled by natural convection while preventing criticality. This can be done by placing trays of suitable shape near the bottom of the vessel, or by reinforcing the bottom of the vessel itself and making provision for cooling it from the outside. Such systems are described in more detail by Lewis (1977), pp. 480 - 493.

It must be emphasised, however, that it is not universally agreed that such devices provide any worthwhile protection, and they have not been fitted on all fast reactors. In order to assess the need for this or any other protective system and the benefit it confers, if any, it is necessary to trace the development of specific accidents and see what influence the protective system has. This is what we turn to in the next section.

5.4 HYPOTHETICAL ACCIDENTS

5.4.1 Accident Sequences

In the previous sections we have identified various inherent features of a fast reactor and various optional protective systems, which between them determine what happens when accidents occur. To ascertain whether the reactor with its protective systems is acceptably safe we have to follow the development of a series of accidents to find out what would actually happen. In this way we can see whether additional protection is needed or indeed whether any of the proposed protective systems are unnecessary. The method is appropriate to reactors of all types, fast or thermal.

It is important to recognise that discussion of accidents has to be conducted in terms of probabilities. There are two reasons for this. Firstly we have to take account of the possibility that a protective system might not work correctly when required. Farmer (1977) shows how this can be done and how estimates can be made of the probability that a given system will work correctly on demand. Secondly we have to take account of the random nature of some of the physical processes involved. The turbulent flow of fluids, for example, or the distribution of defects in a solid body are processes which contain a random element, ultimately at the quantum level. As a result we have to be able to assign probabilities to the possible outcomes of events such as the subjection of a certain structural member to a certain load.

Probabilities can best be incorporated in the discussion by expressing it in "event trees". The method is described by Lewis (1977), pp. 82 - 84, and in more detail in US, NRC (1975). An event tree is a line which traces the development of an accident. At each point where the development can proceed in either of two ways the line branches and probabilities are attached to each branch. As a result all the possible final outcomes can be identified and the probability of each determined.

To make a complete assessment of the safety of a reactor in principle we have to construct an event tree for all the possible events which might initiate an accident. This would be an enormous task, but in practice it is not necessary to go into so much detail because it quickly becomes clear that the risk to the public is dominated by a small number of accidents which can be studied in depth.

All accidents are in some sense unpredictable, because if both the nature of the initiating event and the time of its occurrence were known in advance protective measures would be taken and there would be no accident. Many accident-initiating events are quite predictable in nature, however, and unpredictable only in that it is not known when they will happen. Events such as electrical power failure or fuel element cladding failure fall into this category. The resulting accidents can be analysed quite straightforwardly by event trees.

But it is also necessary to guard against initiating events which are unpredictable in nature as well as in timing: to guard against the accident which has not even been thought of. In principle this is very difficult, but in practice it is possible to make sure that all eventualities have been covered. The method is to make pessimistic assumptions in the event trees. For example we can guard against unforeseen failures of a trip mechanism by assuming arbitrarily that it does not work when required, even though we know of no way this could happen. This is called a "hypothetical accident". In the end a judgement has to be made about how much pessimism of this type should be included in the analysis.

Accident sequences for two initiating events, one concerning a single subassembly and the other the whole core, are discussed below. They are representative of the range of hypothetical accidents which have to be analysed to demonstrate the safety of a given design.

5.4.2 Subassembly Accidents

It is pointed out in section 5.3.3 above that it is necessary to detect a single subassembly accident and prevent it spreading to damage the rest of the core. As an illustration of how the protective systems described in sections 5.3.3 and 5.3.4 do this we shall trace in outline the development of an accident initiated by a partial blockage of the coolant flow in the core region of one subassembly. A complete analysis would be much more detailed than what follows, but would follow these general lines.

The protective systems would of course operate, but for clarity Fig. 5.2 shows the possible sequence of events as it would be if they did not. If no protection were taken there is a possibility that the blockage will cause overheating which might boil the coolant and then melt the fuel, and that this might cause a vapour explosion which in turn might damage the rest of the core. The purpose of the analysis is to turn this sequence of events into an event tree, shown in simplified form in Fig. 5.3, which will show the frequency with which the various possible outcomes can be expected.

First we have to assess the frequency of the initiating event - the occurrence of the blockage. Subassembly inlets are fitted with filters, so a blockage within the subassembly cannot be made of material from outside. The only source of blockage material inside the subassembly would be cladding or fuel from one or more failed fuel elements. It might be supposed that fragments of fuel might come out of a crack in the cladding into the coolant and be trapped at the next support grid downstream (see Fig. 3.16) to form the blockage. It has been shown that this is very unlikely if not impossible. Experiments in which fuel elements have been made to fail in an operating reactor have shown that large failures have not caused blockages (Kramer and others, 1979; Smith and others, 1978). Data like these show that the frequency of the initiating event is very low.

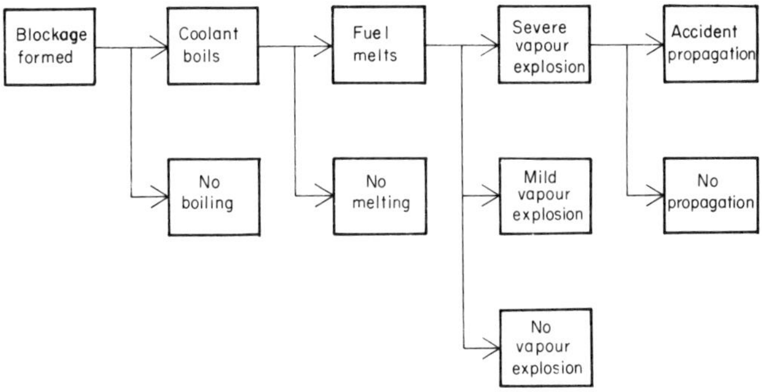

Fig. 5.2. Possible accident sequences for blockage in a single subassembly.

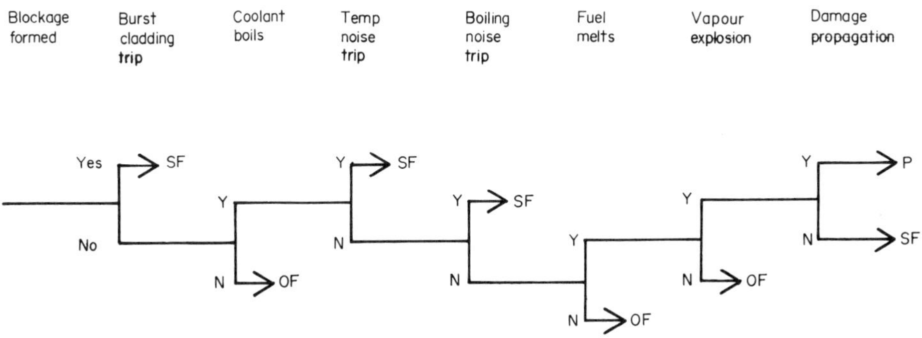

Fig. 5.3. Event tree for subassembly blockage accident showing effects of protective systems.

SF - Shutdown with some failed fuel
OF - Operation with some failed fuel
P - Damage propagated

Since the only source of blockage material is failed fuel a blockage must be accompanied by a substantial failed fuel detection signal, and it can be shown that a blockage can escape detection only if the failed fuel detection system is inoperative. As it may well consist of three separate independent sub-systems this is very unlikely, so the probability of failure (the "no" branch at the first branch point in Fig. 5.3) is very small.

If the "no" branch is taken, the blockage might become large enough to make the coolant boil. (A more detailed analysis would analyse the growth of a blockage, with more branch points in the event tree.) Experiments have determined the temperature distribution in the recirculating wake of a blockage (Kirsch, 1974; Schleisiek, 1974) and typical results are shown in Fig. 5.4. If the blockage is large enough the coolant will boil ("yes" at the second branch point in Fig. 5.3). This is unlikely because the boiling point is some 400 K above the normal coolant temperature, so the blockage has to be quite large.

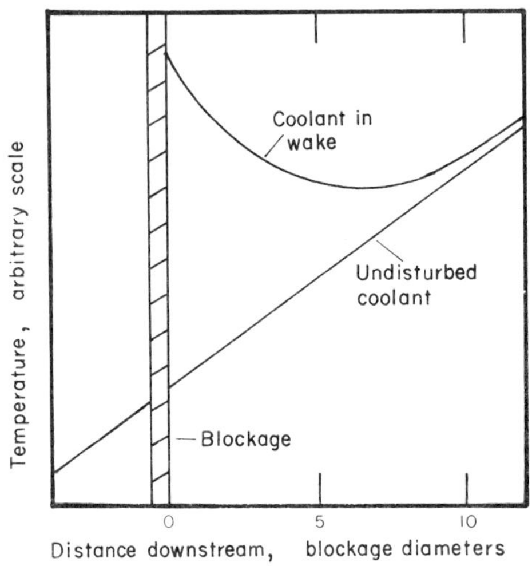

Fig. 5.4. Overheating of the coolant by a blockage.

If the coolant boils it will be detected by the boiling detection system (see section 5.3.4). Only if this fails can the accident progress. Experiments in reactors (Smith and others, 1978) and test rigs (Schleisiek, 1974) have shown that even if the coolant does boil the fuel is cooled adequately by the two-phase coolant. The blockage has to grow still more before the two-phase flow becomes unstable and the surface of the fuel elements dries out. If this lasts for more than a few seconds (which is unlikely) the affected fuel is uncooled and will melt (see section 5.3.4).

The last two branch points in Fig. 5.3 ask whether a vapour explosion takes place and then whether it is violent enough to damage the rest of the core. At present these points have not been resolved experimentally, but the available data (Briggs, Fishlock, and Vaughan, 1979) suggest that the probability of "yes" is low at both branches. It can be shown (Smidt and Schleisiek, 1977) that if a vapour explosion is to be severe enough to disrupt the subassembly wrapper it must involve a large amount of fuel (the amount depends on the design of the wrapper). This in itself makes the probability of "yes" at the last branch point small.

5.4.3 Whole-core Accidents

It is more difficult to trace the development of an accident once it affects the whole of the core. The sequence in which the various events take place is not well determined, and may be different in different parts of the core. As a result it is not so convenient to represent this stage of an accident by an event tree. It is important to remember that the probability of an accident reaching the "whole-core" stage is very small.

The events which could possibly initiate a whole-core accident can be divided into two classes:- "transient over-power", caused by an increase in reactivity, and "loss of flow", caused by a decrease in the coolant flow rate affecting the whole core. The former might be due for example to accidental withdrawal of the control rods or to compaction of the core by the vapour explosion described in section 5.4.2 above; the latter might be the result of a failure of the coolant circulating pumps.

As explained in section 5.3.2 it is very easy to detect such events by a number of independent means. Once detected the defence against a whole-core accident is to insert the absorber rods. As we have seen the rods can be inserted by detaching them from their driving mechanisms, so the shut-down capability is available even if the original accident is caused by a failure of the driving mechanisms. Further assurance can be given by making the absorber rods to different designs, to guard against common-mode failure. Finally the reactivity worth of the rods is chosen so that only a fraction of the total number of rods have to be inserted to shut the reactor down and reduce the power to zero.

Only in the very unlikely event of the shut-down system failing does the accident progress. In a transient over-power accident, overheating of the fuel is likely to lead to melting and clad failure near the centre of the core. Molten fuel and coolant will interact, possibly with a vapour explosion, and the boiling coolant will sweep fuel out of the core, reducing the reactivity and probably terminating the accident. In a loss of flow accident the coolant will boil first near the top of the core where it is hottest, possibly causing dryout, failure of the cladding and release of fission-product gas. Rather later the fuel will melt if the coolant flow rate falls to a very low level. Reactivity changes will be caused by movement of fuel and coolant, and eventually fuel will be swept out of the core and the accident terminated.

The details of what happens in a whole-core accident can be predicted only by a complex computer code. The results of such a calculation depend strongly on the details of the design and the state of the core. The distribution of burnup and fission-product gas in the various subassemblies for example has a strong influence on the course of events. It is not possible to generalise about what happens except to point out that eventually sufficient fuel will be dispersed from the core to render the reactor sub-critical and terminate the accident. The core will be severely damaged and the primary coolant circuit heavily contaminated with fuel and fission products, but two containment boundaries will be intact (section 5.2.1) and the consequences in injury or damage beyond the primary circuit will be small.

5.4.4 Prompt-Critical Excursions

There is one remaining possibility for damage which has to be guarded against. It may be possible under some circumstances (depending on design and on the state of the core) for the movement of fuel and coolant during a whole-core accident to increase the reactivity to prompt critical. In case this turns out to be so, the reactor containment has to be designed to withstand the consequent events.

Paradoxically it is often easier to calculate what happens when prompt criticality is approached than to predict earlier events. At prompt critical the power changes so rapidly that the motion of fuel or coolant is irrelevant and all that matters are the fuel temperature and the resulting pressure which disperses the fuel and terminates the excursion. It is still necessary to use a computer code, but the sequence of events it describes is easier to determine.

When a reactor is close to prompt critical only the prompt neutrons make any significant contribution to the chain reaction and delayed neutrons can be ignored. The prompt neutron lifetime is so short (1 μs or less) that the power level changes essentially instantaneously to keep the total nett reactivity equal to β, the delayed neutron fraction. The nett reactivity is made up of the imposed increase (due to fuel or coolant motion) which is causing the excursion, together with the prompt feedback. Any feedback which depends on heat transfer (such as expansion of the fuel elements which depends on heat being transferred from fuel to cladding) acts too slowly to have any influence. Prompt feedback is provided by the Doppler effect and the expansion of the fuel.

The effect of the feedback is as follows. As the imposed reactivity rises beyond prompt critical the fuel temperature rises with it just fast enough for the Doppler effect to keep the nett reactivity constant. As the fuel temperature rises it has no time to expand or lose heat so its pressure rises. This accelerates the fuel away from the centre of the reactor, which reduces the reactivity and terminates the excursion.

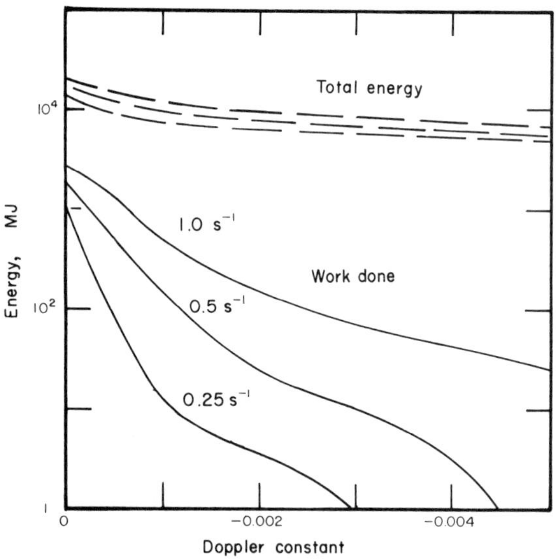

Fig. 5.5. Effect of reactivity ramp rate and Doppler constant on the energy liberated in a prompt-critical excursion.

The amount of energy liberated in such an excursion depends mainly on the Doppler constant (which determines the magnitude of the Doppler feedback, see section 1.6.5) and on the rate at which prompt criticality is approached (usually called the "ramp rate"), as well as on the equation of state of the fuel. Typical results are given by Hicks and Menzies (1965) and Lewis (1977, pp. 245 - 261), and shown in Fig. 5.5. The total energy liberated is more-or-less constant (because pressure rises very steeply with internal energy at constant volume), but the amount of work done by the expanding fuel depends strongly on both the ramp rate and the Doppler constant.

It is the work done in the expansion which determines the load to which the containment is subject. Whole-core accident calculations indicate the prompt-critical ramp rate which should be guarded against, the work done in the resulting excursion is calculated as indicated above, and the primary coolant containment, and especially the reactor vessel and the rotating plugs, is designed to withstand it without more leakage than can be controlled by the last containment boundary, the reactor building.

5.5 RISKS

5.5.1 Accident Consequences

Event trees for hypothetical accidents, carried through to prompt-critical excursions if necessary, will show the amount of radioactive material released to the environment under various assumptions about the operation or failure of the protective systems, together with information about the frequency with which these releases can be expected. It is then necessary to estimate the effect of the possible releases on the surrounding population, and so to determine the risk to which they are subject. The method of doing this which is described here is equally applicable to fast or thermal reactors.

What happens to the material released depends on the weather. The cloud of radioactivity moves with the wind and disperses as it goes. The rate of dispersion depends on the stability of the atmosphere and the nature of the terrain over which the wind blows. Using methods described by Pasquill (1961) it is possible to calculate the exposure to radiation suffered by someone over whom the cloud passes.

The relationship between the exposure which a person experiences and the injury he suffers depends on the composition of the cloud. The most acute danger is usually from ^{131}I, but ^{239}Pu, ^{240}Pu and ^{106}Ru oxides in the form of aerosols, as well as ^{90}Sr and ^{137}Cs, may also cause injury. The susceptibility to injury depends on the age of the person. The injury suffered is in most cases a chance of contracting one or another form of cancer at some later time in life. Sufficient data are available to enable estimates to be made of the number of casualties which will be caused if a given population is subject to a given exposure. This is discussed in detail by Bell (1977).

The result is that for each possible conclusion of each event tree, and for each possible state of the weather, one can calculate N, the expected number of casualties.

5.5.2 Accident Probabilities

The likelihood of an initiating event i can be characterised by a frequency f_i which is the number of times i is expected to occur per reactor year. For example unforeseen failure of a single fuel element might have a frequency f. Formally this means that if F failures occur in n years of operation (of either one reactor or several identical reactors), then $F/n \to f$ as $n \to \infty$.

This frequency might be established experimentally by observing actual fuel element failures, but many accident initiators have much lower frequencies, of 10^{-4} per year, for example, which cannot be justified empirically because it would take millions of years of reactor operating experience to check. Such very low frequencies represent judgements, and are akin to the odds a knowledgeable person would lay that the event in question will take place.

The analysis of the event tree for initiating event i, together with statistical information about the weather, produces estimates of the probability $p_i(N)$ that there will be N casualties given that i happens. The frequency with which accidents of type i cause N casualties is then $f_i p_i(N)$, and the total frequency of accidents causing N casulaties is f(N), where

$$f(N) = \sum_i f_i p_i(N) \qquad 5.1$$

The task of reactor safety analysis is to estimate f(N). It depends on the design of the reactor, the weather experienced at the site, and the distribution of the surrounding population.

5.5.3 Safety Criteria

Whether the reactor is safe enough can be judged by comparing the risk it poses, represented by f(N), with a standard or criterion of acceptability. Such a criterion can be expressed as the frequency a(N) at which accidents causing N casualties are acceptable. The reactor is acceptable if $f(N) < a(N)$ for all N. The use of such criteria and the basis on which they are defined is described by Farmer (1967) to whom the concept is due. A more recent discussion is given by Kinchin (1979).

It is important to remember that f(N) is a scientific and technical statement, but a(N) is not. In the broad sense of the word it is a political statement, because it represents a judgement by society of the risks its members are prepared to accept. It can be based quite legitimately on values which are not scientific.

A criterion of acceptability does not have to be expressed in terms of casualties. It could equally well be put in terms of the monetary cost of the damage caused, and in this way related to the economic benefit brought by the reactor. The setting of the terms in which acceptability is to be judged is also a political, not a scientific, matter.

The legal requirements of reactor safety are not of course expressed in the scientific notation used here. They are usually couched in terms which require interpretation before they can be used to judge the quantitative statement f(N). A curve a(N) then represents an aid to this interpretation. A reactor licensing authority may use it as a means of comparing reactors of different types, susceptible to quite different accidents, in an impartial manner.

A reactor designer may use a curve a(N) as a guide to which of the possible accidents he should guard against. If one gives rise to points above the curve, further protection is needed; if it does not, the protection is adequate.

It is the belief of the author that it is possible to design fast breeder reactors to meet any reasonable standard of acceptability, and more important that if the standard is in accord with the risks posed by and accepted for other enterprises, the reactor will be economically attractive as well. In this way the benefit of the extensive reserves of energy in the world's ^{238}U will be made available for mankind.

References

Anderson, C. A. (1978). Optimization of the Westinghouse/Stone and Webster prototype large breeder reactor. In BNES, Optimisation of Sodium-Cooled Fast Reactors. British Nuclear Energy Society, London. pp. 247-259.

Anderson, R. (1979). Analysis of experimental data on material wastage by sodium-water reaction jets. Nucl. En. 18, 333-342.

Aubert, M., J. M. Chaumont, M. Mougniot, M. Recolin, and M. Acket (1978). Temperature conditions in an LMFBR power plant from primary sodium to steam circuits. In BNES, Optimisation of Sodium-Cooled Fast Reactors. British Nuclear Energy Society, London. pp. 305-310.

Bagley, K. Q., J. L. Bramman, and C. Cawthorne (1971). Fast neutron induced voidage in non-fissile metals and alloys: a review. In BNES, Voids Formed by Irradiation of Reactor Materials. AERE Harwell for British Nuclear Energy Society, London. pp. 1-26.

Bagley, K. Q., J. W. Barnaby, and A. S. Fraser (1973). Irradiation embrittlement of austenitic stainless steels. In BNES, Irradiation Embrittlement and Creep in Fuel Cladding and Core Components. British Nuclear Energy Society, London. pp. 143-153.

Baker, A. R., and R. W. Ross (1963). Comparison of the value of plutonium and uranium isotopes in fast reactors. In USAEC, Breeding Economics and Safety in Large Fast Power Reactors. ANL 6792. pp. 329-365.

Barthold, W. P., C. Tzanos, and J. Beitel (1978). Potential of large heterogeneous reactors. In BNES, Optimisation of Sodium-Cooled Fast Reactors. British Nuclear Energy Society, London. pp. 217-228.

Bell, G. D. (1977). The calculated risk - a safety criterion. In F. R. Farmer (Ed.), Nuclear Reactor Safety. Academic Press, New York.

Board, S. J., R. W. Hall, and R. S. Hall (1975). Detonation of fuel coolant explosions. Nature, 254, 319-321.

Bramman, J. I., C. Brown, J. S. Watkin, C. Cawthorne, E. J. Fulton, P. J. Barton, and E. A. Little (1978). Void swelling and microstructural changes in fuel pin cladding and unstressed specimens irradiated in DFR. In AIME, Radiation Effects in Breeder Reactor Structural Materials. American Institute of Mining Engineers, New York. pp. 479-508.

Briggs, A. J., T. P. Fishlock, and G. J. Vaughan (1979). A review of progress with assessment of MFCI phenomenon in fast reactors following the CSNI specialist meeting in Bournemouth April 1979. In ANS, Fast Reactor Safety Technology, Vol. 3. American Nuclear Society, LaGrange Park, Il. pp. 1502-1511.

Broomfield, A. M., C. F. George, G. Ingram, D. Jakeman, and J. E. Sanders (1969). Measurements of k-infinity, reaction rates and spectra in ZEBRA plutonium lattices. In BNES, The Physics of Fast Reactor Operation and Design. British Nuclear Energy Society, London. pp. 15-24.

Broomfield, A. M., and J. A. Smedley (1979). Operating experience with tube to tubeplate welds in PFR steam generators. In BNES, Welding and Fabrication in Nuclear Industry. British Nuclear Energy Society, London. pp. 3-18.

Brown, C., J. K. Butler, and E. J. Fulton (1979). Void swelling studies of austenitic fuel pin cladding material irradiated in the Dounreay Fast Reactor. In CEA, Irradiation Behaviour of Metallic Materials for Reactor Core Components. Commissariat a l'Energie Atomique, Gif-sur-Yvette. pp. 129-136.

Campbell, R. H. (1973). Primary systems design of sodium-cooled fast reactors. J. Brit. Nucl. En. Soc., 12, 357-365.

Carslaw, H. S., and J. C. Jaeger (1959). Conduction of Heat in Solids, 2nd ed. Clarendon Press, Oxford.

Chamberlain, A., H. A. Taylor, R. H. Allardice, and J. A. Gatley (1978). The optimisation of fuel design in relation to reactor performance and the conflicting demands of other parts of the fuel cycle. In BNES, Optimisation of Sodium-Cooled Fast Reactors. British Nuclear Energy Society, London. pp. 133-6.

Claxton, K. T. (1976). Solubility of oxygen in liquid sodium - effects on interpretation of corrosion data. In ANS, Liquid Metal Technology in Energy Production, Vol. 1. American Nuclear Society, Hinsdale, Ill. pp. 407-414.

Collier, J. G. (1972). Convective Boiling and Condensation. McGraw Hill, New York.

Cox, C. M., and F. J. Homan (1970). Performance analysis of a mixed-oxide LMFBR fuel pin. Nucl. Appl. Tech., 9, 317-325.

Davison, B., and J. B. Sykes (1957). Neutron Transport Theory. Clarendon Press, Oxford.

Duderstadt, J. J., and L. J. Hamilton (1976). Nuclear Reactor Analysis. Wiley, New York.

Dwyer, O. E. (1968). Heat transfer to liquid metals flowing in-line through unbaffled rod bundles: a review. In ASME, Heat Transfer in Rod Bundles. American Society of Mechanical Engineers, New York. pp. 139-168.

Eickhoff, K. G., J. Allen, and C. Boorman (1967). Engineering development for sodium systems. In British Nuclear Energy Society, Fast Breeder Reactors. Pergamon, Oxford. pp. 873-895.

Etherington, E. W., J. J. Bramman, R. S. Nelson, and M. J. Norgett (1975). A UKAEA evaluation of displacement damage models for iron. Nucl. Eng. Des., 33, 82-90.

Evans, P. B. F., E. J. Burton, E. Duncombe, D. Harrison, G. O. Jackson, and N. T. C. McAffer (1967). Control and instrumentation of the Prototype Fast Reactor. In British Nuclear Energy Society, Fast Breeder Reactors. Pergamon, Oxford. pp. 765-782.

Farmer, F. R. (1967). Siting criteria - a new approach. In IAEA, Containment and Siting of Nuclear Power Plants. International Atomic Energy Agency, Vienna. pp. 303-324.

Farmer, F. R. (Ed.) (1977). Nuclear Reactor Safety. Academic Press, New York.

Findlay, J. R. (1974). The composition and chemical state of irradiated oxide reactor fuel material. In IAEA, Behaviour and Chemical State of Irradiated Ceramic Fuels. International Atomic Energy Agency, Vienna. pp. 31-39.

Frame, A. G., W. G. Hutchinson, J. M. Laithwaite, and H. F. Parker (1967). Design of the prototype fast reactor. In British Nuclear Energy Society, Fast Breeder Reactors. Pergamon, Oxford. pp. 291-315.

Friedland, A. J., and C. F. Bonilla (1961). Analytical study of heat transfer rates for parallel flow of liquid metals through tube bundles - II. J. Am. Inst. Chem. Eng., 7, 107-112.

Graham, J. (1971). Fast Reactor Safety. Academic Press, New York.

Greenspan, H., C. N. Kelber, and D. Okrent (1968). Computing Methods in Reactor Physics. Gordon and Breach, New York.

Hans, R., and K. Dumm (1977). Leak detection of steam or water into sodium in steam generators for LMFBRs. Atom. En. Rev., 15, 611-699.

Hayden, O. (1976). Design and construction of past and present steam generators for the UK fast reactors. J. Brit. Nucl. En. Soc., 15, 129-145.

Haywood, R. W. (1975). Analysis of Engineering Cycles, 2nd ed. Pergamon, Oxford.

Henry, R. E. and H. K. Fauske (1975). Energetics of Vapour Explosions. ASME 75-HT-66.

Hicks, E. P. and D. C. Menzies (1965). Theoretical studies on the fast reactor maximum accident. In USAEC, Safety, Fuels and Core Design in Large Fast Power Reactors. ANL 7120. pp. 654-670.

Hoffmann, H., and D. Weinberg (1978). Thermodynamic and fluiddynamic aspects in optimizing the design of fast reactor subassemblies. In BNES, Optimisation of Sodium-Cooled Fast Reactors. British Nuclear Energy Society, London. pp. 133-139.

Horst, K. M. (1978). General Electric/Bechtel prototype large breeder reactor (PLBR). In BNES, Optimisation of Sodium-Cooled Fast Reactors. British Nuclear Energy Society, London. pp. 175-184.

Hummel, H. H., and D. Okrent (1970). Reactivity Coefficients in Large Fast Power Reactors. American Nuclear Society, Hinsdale, Ill.

Kinchin, G. H. (1979). Design criteria, concepts and features important to safety and licensing. In ANS, Fast Reactor Safety Technology, Vol. 1. American Nuclear Society, LaGrange Park, Ill. pp. 1-8.

Kirsch, D. (1974). Investigations on the flow and temperature distribution downstream of local coolant blockages in rod bundle subassemblies. Nucl. Eng. Des., 31, 266-279.

Knowles, J. B. (1976). Principles of nuclear power station control. J. Brit. Nucl. En. Soc., 15, 225-236.

Kramer, W., K. Schleisiek, L. Schmidt, G. Vanmassenhove, and A. Verwimp (1979). In-pile experiments "Mol 7C" related to pin to pin failure propagation. In ANS, Fast Reactor Safety Technology, Vol. 1. American Nuclear Society, LaGrange Park, Ill. pp. 473-482.

Lewins, J. (1978). Nuclear Reactor Kinetics and Control. Pergamon, Oxford.

Lewis, E. E. (1977). Nuclear Power Reactor Safety. Wiley-Interscience, New York.

Lillie, A. F. (1978). Design of the Clinch River breeder reactor steam generators. In IAEA, Design, Construction and Operating Experience of Demonstration LMFBRs. International Atomic Energy Agency, Vienna. pp. 557-571.

Lyon, R. N. (1951). Liquid metal heat transfer coefficients. Chem. Eng. Prog., 47, 75-79.

Martinelli, R.C. (1947). Heat transfer to molten metals. Trans. Am. Soc. Mech. Eng., 69, 947-959.

Meyer, R. O., D. R. O'Boyle, and E. M. Butler (1973). Effect of oxygen-to-metal ratio on plutonium redistribution in irradiated mixed-oxide fuels. J. Nucl. Mat., 47, 265-267.

Mosedale, D., and G. W. Lewthwaite (1974). Irradiation creep in some austenitic stainless steels, Nimonic PE16 alloy, and nickel. In Iron and Steel Institute, Creep Strength in Steel and High-Temperature Alloys. The Metals Society, London. pp. 169-188.

Nettley, P. T., I. P. Bell, K. Q. Bagley, D. R. Harries, A. W. Thorley, and C. Tyzack (1967). Problems in the selection and utilization of materials in sodium cooled fast reactors. In British Nuclear Energy Society, Fast Breeder Reactors. Pergamon, Oxford. pp. 825-849.

Okrent, D. (1961). Performance of large fast power reactors including effects of higher isotopes, recycling and fission products. In IAEA, Physics of Fast and Intermediate Reactors, Vol. 2. International Atomic Energy Agency, Vienna. pp. 271-297.

Okrent, D., K. P. Cohen, and W. B. Loewenstein (1964). Some nuclear and safety considerations in the design of large fast power reactors. In UN, Peaceful Uses of Atomic Energy, Vol. 6. United Nations, New York. pp. 137-148.

Olander, D. R. (1976). Fundamental Aspects of Nuclear Reactor Fuel Elements. Energy Research and Development Administration, Washington.

Pasquill, F. (1961). The estimation of the dispersion of windborne material. Meteor. Mag., 90, (1,063), 33-49.

Perrin, J. S. (1972). Effect of irradiation on creep of UO_2-PuO_2. J. Nucl. Mat., 42, 101-104.

Plein, H. G., R. J. Lipinski, G. A. Carlson, and D. W. Varela (1979). Summary of the first three in-core PAHR molten fuel pool experiments. In ANS, Fast Reactor Safety Technology, Vol. 1. American Nuclear Society, LaGrange Park, Ill. pp. 356-369.

Powell, H. J. (1974). Fission product distribution in fast reactor oxide fuels. In IAEA, Behaviour and Chemical State of Irradiated Ceramic Fuels. International Atomic Energy Agency, Vienna. pp. 379-392.

Schleisiek, K. (1974). Sodium Experiments Relating to Local Coolant Blockages in Test Assemblies Similar to Fuel Assemblies. KFK 1914. (In German).

Smedley, J. A. (1976). Implications of small water leak reactions on sodium heated steam generator design. J. Brit. Nucl. En. Soc., 15, 153-156.

Smidt, D., and K. Schleisiek (1977). Fast breeder safety against propagation of local failures. Nucl. Eng. Des., 40, 393-402.

Smith, C. A., P. A. Simm, and G. Hughes (1979). Analysis of hydride and oxide deposition and resolution in sodium in relation to plugging meter behaviour. Nucl. En., 18, 201-214.

Smith, D. C. G., K. Q. Bagley, C. V. Gregory, G. O. Leet, and D. Tait (1978). DFR special experiments. In IAEA, Design, Construction and Operating Experience of Demonstration LMFBRs. International Atomic Energy Agency, Vienna. pp. 249-260.

Subbotin, V. I., A. K. Papovyants, P. L. Kirillov, and N. N. Ivanovskii (1963). A study of heat transfer to molten sodium in tubes. Sov. J. At. En., 13, 991-994.

Tamplin, L. J. (Ed.), 1963. Reactor Physics Constants, 2nd ed. United States Atomic Energy Commission, Washington.

Tang, Y. S., R. D. Coffield, and R. A. Markley (1978). Thermal Analysis of Liquid-Metal Fast Breeder Reactors. American Nuclear Society, Hinsdale, Ill.

Tattersall, J. O., P. R. P. Bell, and E. Emerson (1967). Large commercial sodium cooled fast reactors. In British Nuclear Energy Society, Fast Breeder Reactors. Pergamon, Oxford. pp. 351-372.

Thorley, A. W., and C. Tyzack (1973). Corrosion and mass transport of steel and nickel alloys in sodium systems. In BNES, Liquid Alkali Metals. British Nuclear Energy Society, London. pp. 257-273.

Tregonning, K., A. Mackay, and K. Buxton (1975). Studies of a mechanism for material wastage by sodium-water reaction jets. J. Brit. Nucl. En. Soc., 14, 77-82.

United States, Nuclear Regulatory Commission (1975). Reactor Safety Study. Appendix 1: Accident definition and use of event trees. WASH 1400, Appendix 1.

Wardleworth, D., and R. C. Wheeler (1974). Reactor physics calculational methods in support of the Prototype Fast Reactor. J. Brit. Nucl. En. Soc., 13, 383-390.

Whittingham, A. C. (1976). An equilibrium and kinetic study of the liquid sodium-hydrogen reaction and its relevance to sodium-water leak detection in LMFBR systems. J. Nucl. Mat., 60, 119-131.

Yiftah, S. (1961). Effect of the plutonium isotopic composition on the performance of fast reactors. In IAEA, Physics of Fast and Intermediate Reactors, Vol. 2. International Atomic Energy Agency, Vienna. pp. 257-270.

Index

Printed in United States of America by Edwards Brothers, Inc.